Geobotany Studies

Basics, Methods and Case Studies

About the Series

The series includes outstanding monographs and collections of papers on a range of topics in the following fields: Phytogeography, Phytosociology, Plant Community Ecology, Biocoenology,Vegetation Science, Eco-informatics, Landscape Ecology, Vegetation Mapping, Plant Conservation Biology, and Plant Diversity. Contributions should reflect the latest theoretical and methodological developments or present new applications on large spatial or temporal scales that will reinforce our understanding of ecological processes acting at the phytocoenosis and vegetation land scape level. Case studies based on large data sets are also considered, provided they support habitat classification refinement, plant diversity conservation orvegetation change prediction. ***Geobotany Studies: Basics, Methods and Case Studies*** is the successor to *Braun-Blanquetia*, a journal published by the University of Camerino from 1984 to 2011 incooperation with the *Station Internationale de Phytosociologie* (Bailleul, France) and the *Dipartimento di Botanica ed Ecologia* (Universitàdi Camerino, Italy) and under the aegis of the *Société Amicale Francophone de Phytosociologie*, the *Société Française de Phytosociologie*, the *Rheinold-Tüxen-Gesellschaft* and the *Eastern Alpine and Dinaric Society for Vegetation Ecology*. This series promotes the expansion, evolution, and application of the invaluable scientific legacy of the Braun-Blanquet school.

More information about this series at http://www.springer.com/series/10526

Franco Pedrotti
Editor

Climate Gradients and Biodiversity in Mountains of Italy

Springer

Editor
Franco Pedrotti
Department of Botany and Ecology
University of Camerino
Camerino, Italy

ISSN 2198-2562 ISSN 2198-2570 (electronic)
Geobotany Studies
ISBN 978-3-319-67966-2 ISBN 978-3-319-67967-9 (eBook)
https://doi.org/10.1007/978-3-319-67967-9

Library of Congress Control Number: 2017959577

Printed on acid-free paper

This Springer imprint is published by Springer Nature
The registered company is Springer International Publishing AG
The registered company address is: Gewerbestrasse 11, 6330 Cham, Switzerland

Preface

This volume is dedicated to the examination and interpretation of climatic gradients and plant biodiversity of some mountain systems of Italy, with a contribution that also takes into consideration the biodiversity of the avian fauna.

The Italian peninsula extends between the latitudes of 36°N and 47°N and belongs to two vegetation zones: the Mediterranean and the temperate (Rivas Martìnez, *Bioclimatic Map of Europe*, 1996). The Mediterranean zone is characterized by the evergreen sclerophyllous associations of the *Quercetea ilicis* class. The temperate zone comprises deciduous broadleaf associations of the *Carpino-Fagetea, Quercetea pubescentis*, and *Quercetea roboris* classes, plus the needle-leaved *Vaccinio-Piceetea, Pyrolo-Pinetea*, and *Erico-Pinetea* classes (latitudinal zonation of the vegetation).

This territory is characterized by mountain systems that in part are found in the temperate zone (the Alps and the north-central Apennines) and in part in the Mediterranean zone (the southern Apennines and the mountains of Sicily and Sardinia). Every mountain system, in turn, is characterized by altitude belts with their own plant associations, according to the altitude above sea level (altitudinal zonation of the vegetation). The altitudinal zonation is very different according to the geographic position of the mountain system in the Mediterranean zone or the temperate one (Blasi, *La vegetazione d'Italia*, 2010; Pedrotti, *Plant and vegetation mapping*, Springer, 2013).

A number of well-known general contributions examine the plant biodiversity of entire mountain groups, for example, the Alps and Apennines (Pedrotti and Gafta, *The High Mountain Flora and Vegetation of the Apennines and the Italian Alps*, Springer, 2003), the Dolomites (Pignatti and Pignatti, *Plant Life of Dolomites*, Springer, 2014), and the mountains of Sicily (Raimondo, *Carta del paesaggio vegetale e della biodiversità vegetale della Provincia di Palermo*, University of Palermo, 2000).

This volume gathers case studies in plant diversity from some representative mountain systems in the Mediterranean and temperate zones of Italy.

Biodiversity was evaluated on the basis of the number of species (*specific biodiversity*), plant communities (*coenotic biodiversity*), and vegetation series (*landscape diversity*), according to the definition of Géhu (*Dictionnaire de Sociologie et Synécologie végétales,* Cramer, 2006).

For the Alps, Wilhalm and Prosser examine the species biodiversity (also with the help of very detailed location maps) of the sector of the central Alps that corresponds to the basin of the Adige, with some nearby valleys, between the watershed to the north and the Prealps to the south (Alto Adige and Trentino). Pedrotti investigates the vegetation series of the same territory, in relation to the three climatic sectors identified: prealpine, alpine and endoalpine. Aleffi explores the relationships between the distribution of some species of bryophytes and the main mesoclimatic gradients along a transect through the Valle dell'Adige between 46°40′N and 45°42′N. Siniscalco studies the ways alien species invade the western Alps, which to date are little affected by this phenomenon, unlike the plains and hills.

For the Apennines, Ferrari examines the tree line and the biodiversity of the vegetation of the northern Apennines.

For the mountains of Sicily, Bazan conducts a diachronic analysis of the beech forests of the Monti Nebrodi.

The contribution of Venanzoni interprets the chorology of associations of *Magnocaricetalia* order throughout Italy, relating it to the climatic and geographic gradients. In all, he describes 55 associations, reporting for each of them the distribution in the temperate zone (distinguished into the alpine and continental) and the Mediterranean zone.

Cianfaglione presents the Signal Project Italian site. This project investigates the effects of extreme weather events on secondary grassland and the role of selected alien species, mowing, biodiversity, productivity, and functional traits, in Italy and along a European gradient, sharing data with other partners.

For the Marche Region, Forconi describes the biodiversity of the avian fauna in relation to the altitudinal gradient and the potential vegetation.

The approaches used in the various contributions are based on fairly different methodologies but all center on analysis of the biodiversity in function of climatic gradients or, in some cases, also on other factors.

Camerino, Italy Franco Pedrotti

Contents

Floristic Biodiversity in South Tyrol (Alto Adige) 1

Thomas Wilhalm

Abstract

An overview is provided of the knowledge regarding the vascular flora of Alto Adige or South Tyrol, Italy's northernmost province, gained through its thorough study during the last decades, primarily by means of comprehensive floristic mapping.

The flora of South Tyrol comprises some 2600 taxa, 83% of which are classified as indigenous or archaeophytic. Of the remaining 17% alien species, 7% are considered as naturalized and 10% as casual. A detailed analysis of the composition of the flora reveals a huge proportion of species with a pan-European distribution range, of species with the main area of distribution in the nemoral zone, or in the zone of European high mountains (alpid), or in the sub-Mediterranean region, and of Alpine species. This reflects the geographic location of the region both in the southern part of Central Europe and in the midst of the Alps.

One phytogeographical particularity is the relatively high proportion of arctic-alpine, boreal and pontic elements, which is due to postglacial migrations and associated area disjunctions and, in the case of pontic elements, to particularly favorable microclimates in parts of South Tyrol.

Keywords

Vascular flora • Floristic biodiversity • South Tyrol • Alps • Phytogeographical elements • Chorology

T. Wilhalm (✉)
Museum of Nature South Tyrol, Via Bottai 1, 39100 Bolzano, Italy
e-mail: thomas.wilhalm@naturmuseum.it

F. Pedrotti (ed.), *Climate Gradients and Biodiversity in Mountains of Italy*,
Geobotany Studies, https://doi.org/10.1007/978-3-319-67967-9_1

1.1 Introduction

Along with topography, geology and climate, the Flora is among the most important parameters to characterize a region and has thus been an important subject of interest since the beginning of modern regional studies.

Although it is a high mountain area and poorly accessible at the time, Alto Adige or South Tyrol, i.e. the Province of Bolzano, was already an important destination for floristic explorations 200 years ago. Around 1900, the "Old Tyrol", i.e. the current provinces of Bolzano and Trento in Italy and the state of Tirol in Austria, was among the best floristically investigated regions of Europe.

The aim of the present article is to synthesize the status quo of the floristics in South Tyrol and to analyze the diversity and composition of its vascular flora.

1.2 Study Area

1.2.1 Topography

South Tyrol is the northernmost province of Italy. It is situated in the center of the Alps and covers an area of 7400 km^2. As a high mountain region, the relief is very pronounced, including the deep and mostly wide main valleys of the rivers Adige, Isarco and Rienza, numerous lateral valleys including many that are high-altitude valleys, as well as various mountain ranges. Among the latter is the main ridge of the Alps which constitutes an important watershed as well as the border with Austria and which includes great massifs such as the mountains of the Ötztal and the Zillertal valleys. Other important mountain chains marking the typical South Tyrolean landscape are the Dolomites in the southeast and the Ortles range in the southwest. The height difference of the region's relief is striking, with the lowest elevation at 200 m being the Salorno ravine in the Adige Valley and the highest at 3905 m a.s.l., the summit of Mt. Ortles.

The mountainous character of South Tyrol is also reflected by the fact that 60% of the area is situated above 1600 m and that 21% of it belongs to the alpine and nival zone (Autonome Provinz Bozen-Südtirol 2010).

1.2.2 Geology

The tectonic units which characterize the geological landscape of South Tyrol are the "Austroalpin", the "Southern Alps", and the "Penninic" (Fig. 1.1). The first is comprised for the most part of acid cristalline rocks, which is reflected by a comparatively poor flora. To a minor degree, there are also areas of basic Triassic sediments, which are areas of elevated floristic biodiversity (see below). In the Southern Alps, there are more rocks of different chemical composition which play a key role in determining the nature of soils and thus the composition of the flora.

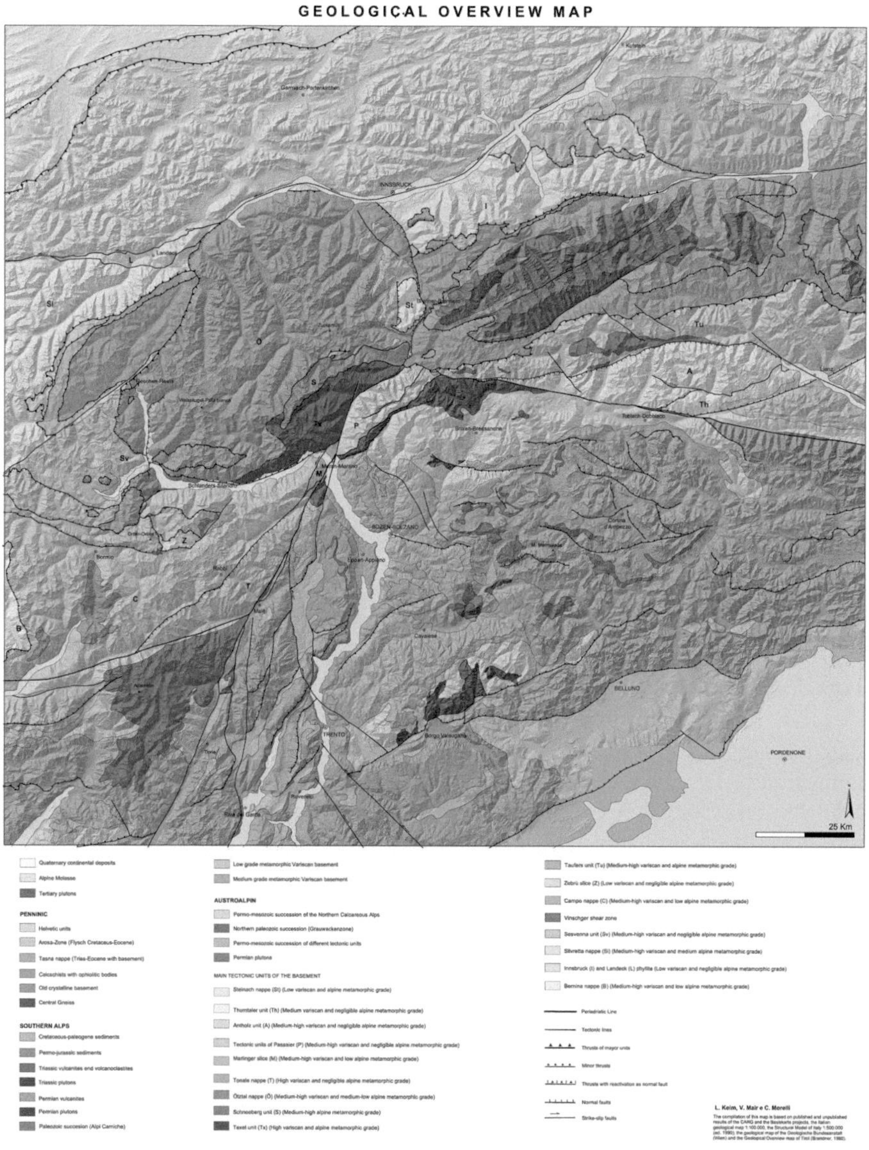

Fig. 1.1 Geological sketch map of South Tyrol showing the most important tectonic units. See text for more details. Source: Geological Office of the Autonomous Province of Bolzano—South Tyrol (with friendly permission of the editor V. Mair)

Particularly noteworthy are the basic Triassic sediments of which the famous Dolomites are composed, as well as the porphyry zone around Bolzano. The Penninic has a small portion of the territory, namely in the north-central part. Nevertheless, the predominating calcareous schists give rise to an extraordinarily rich alpine flora (see below).

1.2.3 Climate

Although located in the midst of the Alps, South Tyrol enjoys particular climatic conditions with relatively low annual precipitation (700–800 mm at lower altitudes, 450–500 mm in the inner Alpine dry valley of Venosta) (Fig. 1.2) and both high solar radiation and a high number of solar days. The climatic insular effect is due first of all to the region's location south of the main ridge of the Alps which to a high degree shields from Atlantic depressions that affect the weather, particularly in winter. Secondly, the deep Adige Valley is open toward the south, allowing warm Mediterranean air to penetrate deep into the interior of the Alps. This latter effect is decisive in weakening the climate with an otherwise subcontinental touch that characterizes the main area of South Tyrol and becomes important in the southern part of the province. In addition to the low precipitation and the low mean annual temperature including winter frosts throughout the entire area of the region area, the continentality of the climate is expressed by the distribution of precipitation. Indeed, the least precipitation is observed between December and April while most precipitation occurs in the form of rainfall during thunderstorms in the summer months, typically in connection with Mediterranean cyclones ("Genova-depression") (Fig. 1.2).

In detail, South Tyrol covers a series of climate groups due to different altitudes. According to the definitions provided by Kottek et al. (2006), the major part of the settled area belongs to the temperate humid climate (group C), while the Upper Puster Valley must be assigned to the boreal climate (group D) and the high mountain areas to the snow climate (group E). One particularity in Central Europe is the climate of the Venosta Valley, which in part approaches the steppe climate (group BS). There are only two regions in all of the Alps that share these conditions: Aosta (Northwestern Italy) and Valais (Switzerland).

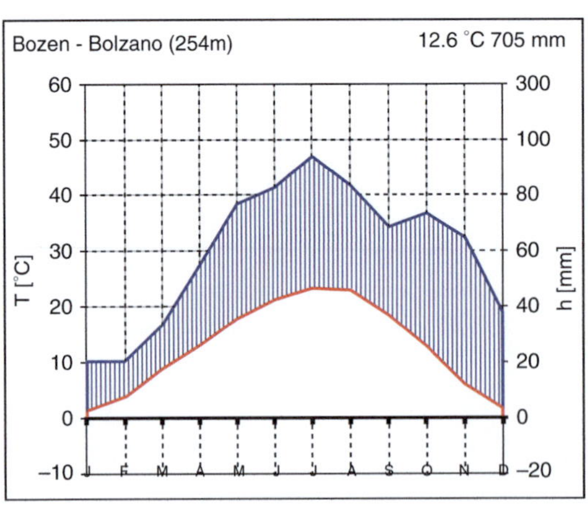

Month	Temperature [°C]	Precipitation [mm]
Jan	1.2	19.7
Feb	3.9	20.2
Mar	9.1	33.0
Apr	13.0	53.1
May	17.7	76.9
June	21.1	82.3
July	23.4	93.7
Aug	22.9	83.7
Sep	18.4	68.7
Oct	12.8	72.8
Nov	6.0	64.7
Dec	1.7	36.6

Mean annual temperature: 12.6 °C

Accumulated precipitation: 705 mm

Fig. 1.2 Climate chart (*left*) and mean monthly temperature and precipitation (*right*) of Bolzano, capital of Alto Adige, for the period 1981–2010. Source: Hydrographic Office of the Autonomous Province of Bolzano—South Tyrol (kindly provided by Dieter Peterlin)

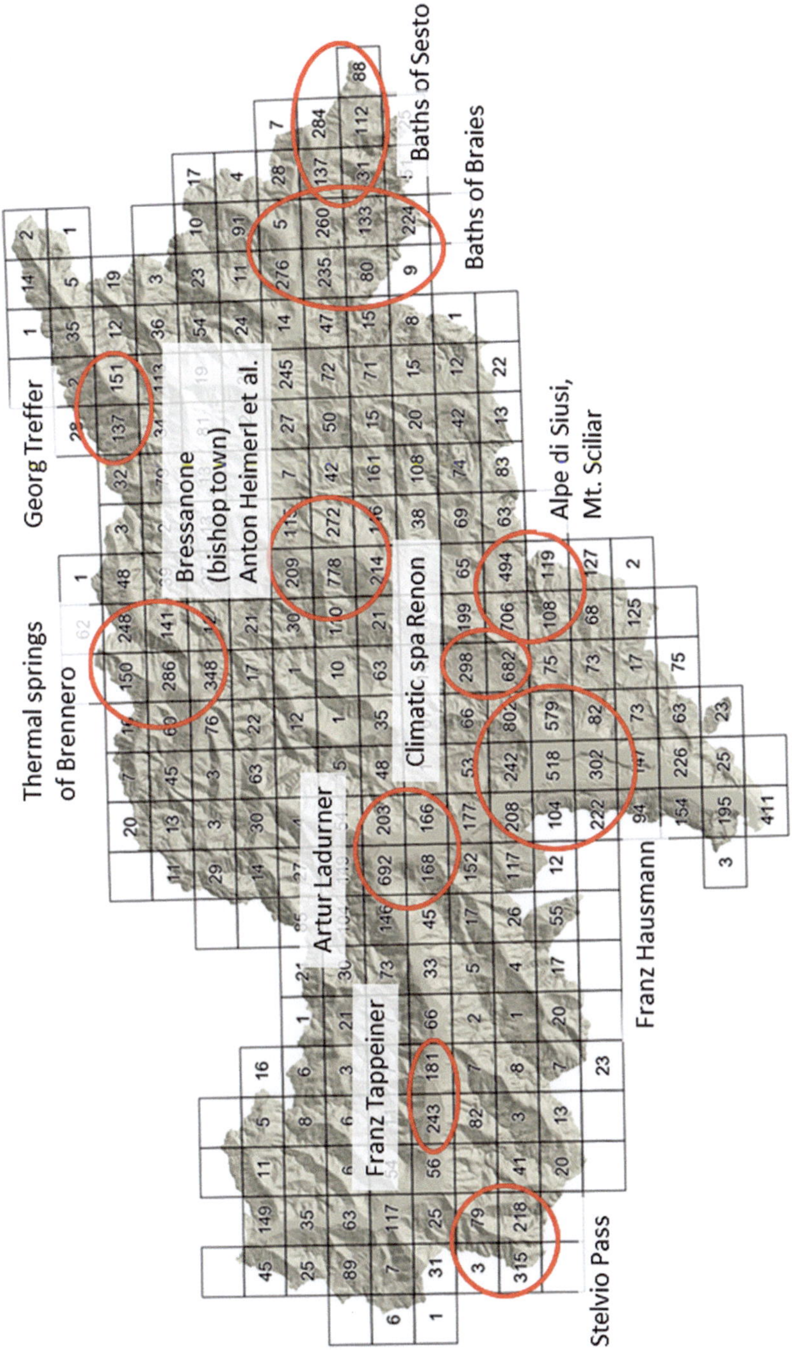

Fig. 1.3 The vascular flora of South Tyrol: Number of taxa accepted as valid today per operational geographic unit (*grid square*, i.e. "quadrant" according to Niklfeld 1971) known in 1913 (data source: Dalla Torre and Sarnthein 1906–1913). The *red circles* indicate centers of enhanced floristic activities, the corresponding captions the obvious reason for that (famous, well visited localities or domains of important collectors)

1.3 Notes on the History of Floristics in South Tyrol

The outstanding floristic activities in South Tyrol in the nineteenth century culminated in the monumental flora published by Dalla Torre and Sarnthein (1906–1913). When looking at the species number per operational geographic unit known at that time (Fig. 1.3), one realizes that the overall floristic exploration was remarkably good. It is unmistakable, however, that there were clear centers of enhanced activity while other parts of the territory were seriously neglected. The areas of elevated species number obviously coincide with the domains of important collectors or with points of attraction, such as thermal baths or famous mountain passes (Fig. 1.3).

The systematic and area-wide inventory of the vascular plant flora of South Tyrol begins with the floristic mapping-project at the end of the 1970s with a first peak reached by the field work done by employees of the Botanical Institute of the University of Vienna in the period 1980–1998. From 1998 on the newly founded Museum of Nature South Tyrol takes over the operational coordination and execution of mapping. The remaining grid cells were mapped (Fig. 1.4), new findings published (e.g. Niklfeld 2003; Wilhalm et al. 2003) and efforts made to an up-to-date synthesis of the vascular flora including the compilation of an annotated checklist (Wilhalm et al. 2006) and a Red List of endangered species (Wilhalm and Hilpold 2006). For more details regarding floristics in South Tyrol in the twentieth century, see Wilhalm (2012).

1.4 Results of the Current Research on the Vascular Flora

1.4.1 Distribution Data

One of the most important result of floristic mapping in the period 1980–2013 is the collection of comprehensive, area-wide data on the distribution of all vascular plant taxa occurring in South Tyrol. The processing stage of the single grid cell ("quadrant", see Fig. 1.4) may be regarded as good to very good, although some (remote) parts of the territory remain to be explored in more detail.

Even though a detailed analysis of the flora is still missing, the floristic biodiversity and its distribution over the territory of South Tyrol may be assessed by simply looking at the number of taxa ascertained in each grid cell (Fig. 1.4). By doing so, two lowland regions with grid cells covering areas mostly below 1500 m and three mountain regions with areas above 1500 m can be identified which demonstrate comparatively high species numbers. Among the lowland regions are the Adige Valley between Salorno and Merano and the basin of Bressanone where more than 1000 taxa are observed in single grid cells. Of course, in these regions, the proportion of neophytic and ephemerous species assumes particular importance in comparison to the mountain regions. The three mountain regions with a particularly high floristic biodiversity are the mountains of Sciliar, the mountains of Brennero and Fundres as well as the southern part of the Sesvenna and the western part of the Ortles massif (Fig. 1.4).

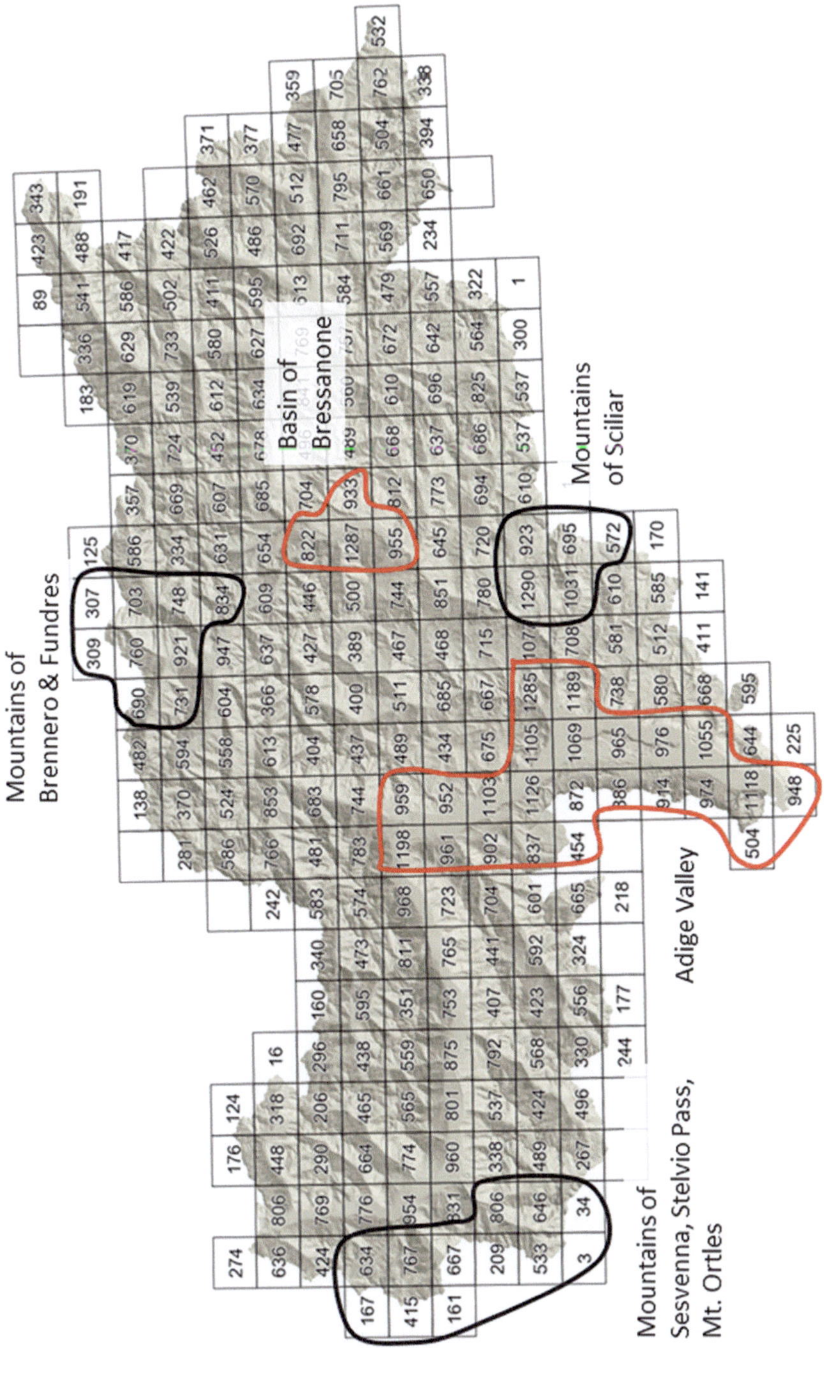

Fig. 1.4 The vascular flora of South Tyrol: Numbers of taxa reported per grid cell ("quadrant") as a result of the current floristic mapping (source: database of the Museum of Nature South Tyrol, Bolzano, September 2013). The *circles* mark areas with high floristic diversity in lowland regions (*red*) and mountain regions (*black*)

These centers of high diversity of mountain flora essentially correspond to areas with predominating calcareous rocks, such as limestone, dolomite, and calcareous schists, in combination with acid rocks, such as gneiss, mica-schist, porphyry, and with volcanic rocks of different types. On the other hand, areas such as great parts of the Ötztal Alps and the Sarntal Alps with predominating gneiss and porphyry, respectively, are noticeably species-poor (Fig. 1.4).

1.4.2 Species Number

A remarkable increase of species number can be shown when comparing the number reported in the two historical floras (Hausmann 1851–1854 and Dalla Torre and Sarnthein 1906–1913) and the most recent checklist (Wilhalm et al. 2006) (Fig. 1.5). The increment in the second half of the nineteenth century can easily be attributed to the enforced floristics in terms of area covered and to the overall increased activity in taxonomy in that period. The increased number of taxa in the current checklist compared to that of the hitherto most comprehensive flora published by Dalla Torre and Sarnthein (1906–1913) is primarily due to the rapid rise of adventitious species and the establishment of neophytes in the course of the twentieth century. The number of newly arrived allochthonous species exceeds, by far, that of species missing or extinct since 1900. Among autochthonous species, only few were discovered after Dalla Torre and Sarnthein's flora (see e.g. Wilhalm et al. 2006), while a couple of new ones resulted from new taxonomic concepts and taxa that were formerly not recognized as separate species. According to Wilhalm et al. (2006), the vascular flora of South Tyrol comprises 2579 taxa of which 2169 are classified as indigenous or archaeophytic, 134 as established neophytes, 79 as possibly naturalized, and 185 as ephemerous. For 12 taxa, the status, whether autochthonous or neophytic, is uncertain. It is clear that these numbers are not stable and that especially the number of allochthonous species increases continuously (e.g. Wilhalm et al. 2005, 2009; Wilhalm 2011).

Fig. 1.5 Number of taxa known in South Tyrol accepted as valid today, reported in Hausmann (1851–1854), Dalla Torre and Sarnthein (1906–1913) and Wilhalm et al. 2006 (from: Wilhalm 2012)

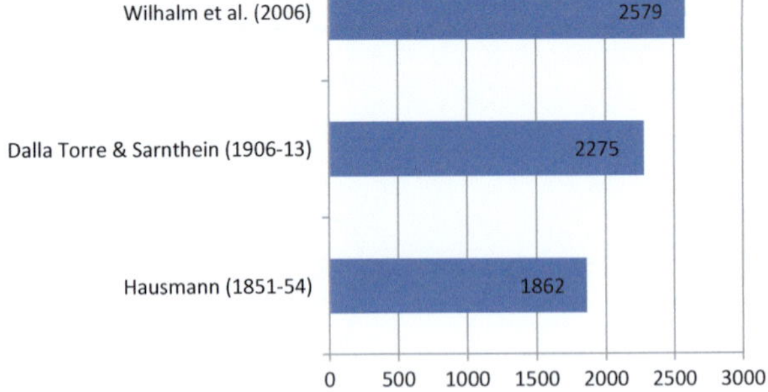

1.4.3 Composition of the Flora

1.4.3.1 Chorotypes

The flora of a determined territory can be well described by analyzing the composition of the (indigenous) plant taxa in view of phytogeographical aspects, i.e. of their main distribution area. In South Tyrol, the greatest portion is accounted for by species with a pan-European distribution range and those with the main area of distribution in the nemoral zone, or in the zone of european high mountains (alpid), as well as strict Alpine species. Also, the number of sub-Mediterranean elements shows a remarkable proportion (Fig. 1.6). This all reflects the geographic location of the region both in the southern part of Central Europe and in the middle of the Alps. Its location in the geographical center of the Alps, on the other hand, causes the presence of both western alpine and eastern alpine elements (Figs. 1.6 and 1.7). The fact, that the proportion of eastern alpine elements is comparatively higher indicates that South Tyrol belongs phytogeographically to the Eastern Alps.

Of particular phytogeographical interest are the arctic-alpine, boreal and pontic elements, which play a significant role in the flora of South Tyrol (Fig. 1.6). These elements have been established in the course of postglacial migrations and associated area disjunctions. As to the boreal elements, special mention must be made of species like *Andromeda polifolia*, *Botrychium lanceolatum*, *Carex heleonastes*, *Carex capitata*, *Carex chordorrhiza*, *Carex vaginata*, *Corydalis capnoides*, *Ranunculus reptans*, *Vaccinium microcarpum*, and *Woodsia ilvensis*, all of which are species which in the Alps are found only in small and very isolated populations and which, also because of the fact that they mostly grow in wet habitats, are generally classified in red lists as very endangered (e.g. Conti et al. 1992; Prosser 2001; Wilhalm and Hilpold 2006). Other important boreal elements are *Linnaea borealis* and *Trientalis europaea* (Fig. 1.8). The South Tyrolean populations, particularly those of *L. borealis*, are the largest throughout Italy.

Among the arctic-alpine element, representatives which are typical although rather rare in the Alps are: *Astragalus frigidus*, *Carex bicolor*, *Carex maritima*, *Carex microglochin*, *Carex norvegica*, *Festuca vivipara*, *Juncus arcticus* (Fig. 1.9), *Minuartia biflora* (Fig. 1.9), *Oxytropis lapponica*, *Pedicularis oederi* (Fig. 1.9), *Potentilla nivea* (Fig. 1.9), *Ranunculus pygmaeus*, *Saxifraga cernua*, *Thalictrum alpinum*, and *Woodsia alpina*.

Due to favorable microclimates, it was possible for pontic species to establish themselves after postglacial migration from their core distribution area of Russia into the driest parts of the Alps, the so-called "inner Alpine dry valleys", i.e. Valais (Switzerland), Aosta (Northwestern Italy) and the Vinschgau Valley (South Tyrol). These include, among others, the following species, some of which (*) in Italy are (today) found only in South Tyrol: *Carex stenophylla*, *Carex supina** (Fig. 1.10), *Dracocephalum austriacum*, *Festuca rupicola*, *Festuca valesiaca* (Fig. 1.10), *Fumaria schleicheri*, *Oxytropis pilosa*, *Seseli pallasii*, *Stipa capillata*, and *Veronica dillenii*.

Beside some agamospermic micro-species (e.g. *Hieracium venostorum*), there are no vascular plant species strictly endemic to South Tyrol. The obvious reason

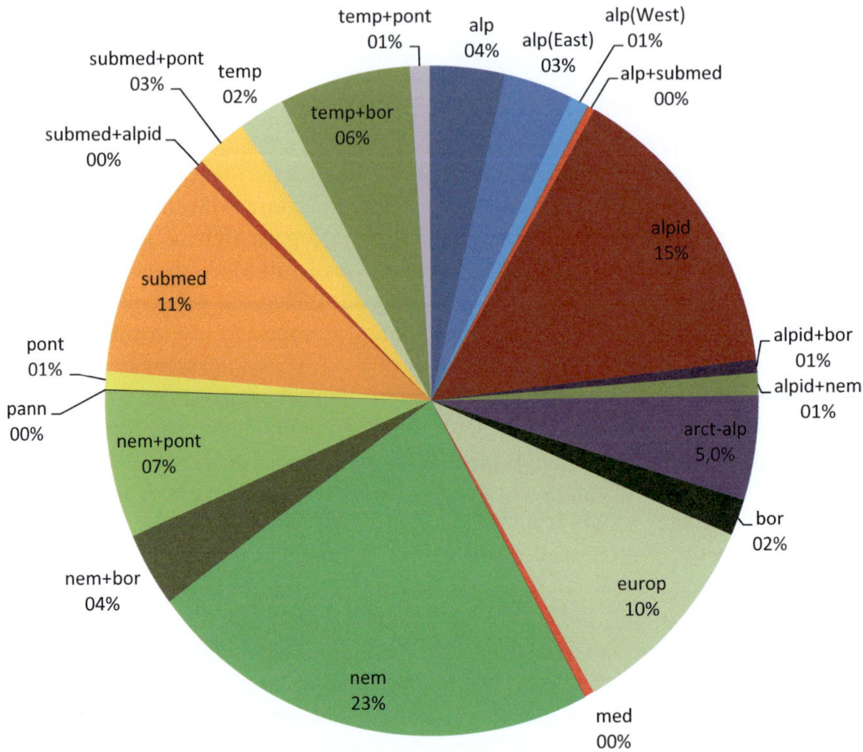

Fig. 1.6 The vascular flora of South Tyrol: Composition of the species (percentage) according to their main distribution (in Europe). Only indigenous and archeophytic species are considered excluding agamospermic taxa, in total 2030
Legend of the chorotypes used in the figure (after Schneeweiß and Niklfeld in prep., partly simplified):
alp = alpine: main distribution in the Alps;
alpid: distributed in the high mountains of Europe;
arct-alp = arctic-alpine: main distribution in arctic regions and in the Alps;
bor = boreal: main distribution in the coniferous forest area of nordic regions (of Eurasia and North America);
europ = European: distributed throughout Europe;
med = Mediterranean: main distribution in the Mediterranean region;
nem = nemoral: main distribution in the European zone of deciduous forests;
pann = pannonic: main distribution in the pannonic region (Eastern Austria and Hungaria);
pont = pontic: main distribution in the pontic region, i.e. the steppe-area north and northeast of the Black Sea;
submed = sub-Mediterranean: main distribution in the peripheral areas of the Mediterranean;
temp = temperate: main distribution in the temperate climate zones of Europe

for this is that because of the central Alpine position, almost no *massifs de refuge* are likely to have been available for species survival in the territory of South Tyrol during the last ice age. Nevertheless, some of Alpine stenoendemites also cover a notable area in South Tyrol, namely *Braya alpina*, *Draba dolomitica*, *Rhizobotrya*

Fig. 1.7 The vascular flora of South Tyrol: Examples of Alpine species and their distribution in South Tyrol. *Above*: *Achillea nana* (West-Alpine, all photos by T. Wilhalm except where otherwise indicated). *Below*: *Pedicularis aspleniifolia* (East-Alpine). A *circle* indicates the presence in the corresponding grid cell (quadrant); *red* = last record after 1980, *orange* = last record between 1913 and 1979, *yellow* = last record before 1913

alpina, *Saxifraga facchinii* (Fig. 1.11), and *Sempervivum dolomiticum*. On the other hand, the list of species endemic to a larger part of the Alps including South Tyrol is quite long; it includes primarily species endemic to the southeastern Alps, such as *Achillea oxyloba*, *Androsace vitaliana* subsp. *sesleri*, *Androsace hausmannii*, *Aquilegia einseleana*, *Arenaria huteri*, *Campanula carnica*, *Campanula morettiana*, *Draba dolomitica*, *Festuca alpestris*, *Galium baldense*, *Galium margaritaceum*, *Gentiana terglouensis*, *Gentianella engadinensis*, *Minuartia*

Fig. 1.8 The vascular flora of South Tyrol: Examples of boreal species and their distribution in South Tyrol. *Above*: *Linnaea borealis*. *Below*: *Trientalis europaea* (photo by E. Schneider-Fürchau). For explanation of symbols see legend in Fig. 1.7

cherlerioides subsp. *cherlerioides*, *Paederota bonarota*, *Pedicularis elongata*, *Physoplexis comosa*, *Phyteuma sieberi*, *Primula daonensis*, *Saxifraga hostii* subsp. *rhaetica*, *Saxifraga squarrosa*, and *Saxifraga tombeanensis*. Their occurrence may be explained by a gradual (re-)immigration from *massifs de refuge* at the southern edge of the Alps during the post glacial area.

Fig. 1.9 The vascular flora of South Tyrol: Examples of arctic-alpine species and their distribution in South Tyrol. From *top* to *bottom*: *Juncus arcticus*, *Minuartia biflora* (photo by J. Winkler), *Pedicularis oederi*, *Potentilla nivea*. For explanation of symbols see legend in Fig. 1.7

Fig. 1.9 (continued)

1.4.3.2 Chorological Status

More than four fifths of the flora of South Tyrol is classified as autochthonous, i.e. either indigenous or archeophytic, while the remainder (17%) is covered by species alien to the flora (Fig. 1.12). It is noteworthy that among the alien species, only 7% of the total are considered as naturalized (Figs. 1.12 and 1.13), compared to the remaining 10% which are casual. Some alien species have been established within the indigenous vegetation over the last decades and centuries in a rather inconspicuous manner so as to appear today as part of the indigenous flora (e.g. *Geranium pyrenaicum*, *Matricaria chamomilla*). As for the so-called "problematic species", only very few have exerted a serious impact on the environment thus far, above all those which significantly alter the structure and ecology of the indigenous vegetation. Such "transformers" include the following species: *Ailanthus altissimus*, *Buddleja davidii*, *Impatiens glandulifera*, *Impatiens parviflora*, *Robinia pseudacacia*, *Solidago canadensis*, *Solidago gigantea*, and *Senecio*

Fig. 1.10 The vascular flora of South Tyrol: Examples of pontic species and their distribution in South Tyrol. *Above*: *Carex supina*. *Below*: *Festuca valesiaca*. For explanation of symbols see legend in Fig. 1.7

Fig. 1.11 The vascular flora of South Tyrol: Example of a stenoendemic species subendemic to South Tyrol and its distribution in South Tyrol: *Saxifraga facchinii*. For explanation of symbols see legend in Fig. 1.7

inaequidens. Species causing a potential health risk, such as *Ambrosia artemisiifolia* and *Heracleum mantegazzianum*, still play a very minor role in South Tyrol. The same holds true for species which represent a potential threat to agriculture (e.g. *Amaranthus* spp.).

For additional details see Celesti-Grapow et al. (2009).

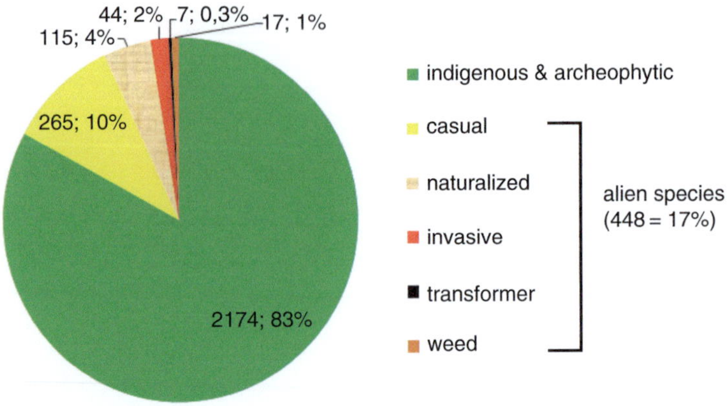

Fig. 1.12 Composition of the vascular flora of South Tyrol: chorological status. Year of reference: 2006. Source: Wilhalm et al. 2006. Categories from Pysek et al. (2004)

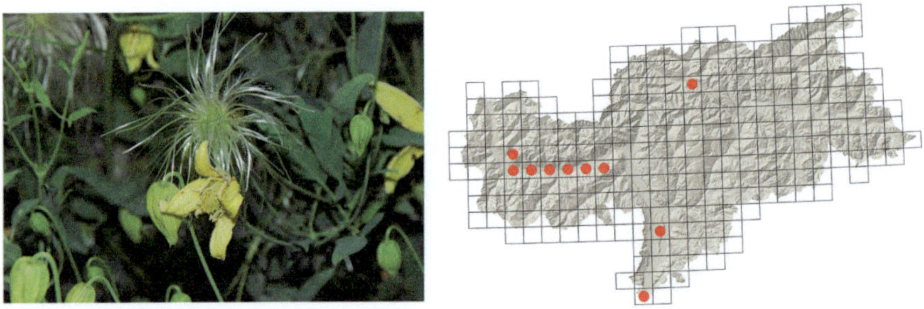

Fig. 1.13 The vascular flora of South Tyrol: *Clematis tangutica*. Example of a naturalized species and its distribution in South Tyrol. This species of Chinese-Mongolian origin is cultivated as an ornamental plant and was observed as naturalized the first time in Italy in the Venosta Valley (Wilhalm et al. 2003). For explanation of symbols, see legend in Fig. 1.7

1.4.4 Threats to the Floristic Biodiversity in Alto Adige

The floristic biodiversity of South Tyrol, though (still) high, is nevertheless endangered. This holds true first and foremost for the valley floors. Wilhalm and Hilpold (2006) recognized that the main factor threatening the vascular flora is the intensification of agriculture, expressed by the eutrophication of entire areas including nutrient-poor grasslands, wetlands, aquatic habitats, etc. Other important factors are the abandonment of traditional land use and settlement activities (urbanization). One factor which may be unique throughout the entire Alps, causing a marked decline in floristic biodiversity, is the continuing development of mountain areas. In fact, South Tyrol claims to be the only province in the Alps with almost no migration from mountain regions thanks to good infrastructures. The downside of this reality: the more remote mountain regions are easily accessible, the more farming is intensified in close vicinity to the farming houses and the more grassland, formerly cultivated extensively, is abandoned at a greater distance from the settlements for reasons of rationalization. The negative implications for the flora are clear: mountain meadows and pastures, hitherto a precious pool of species and genetic diversity, are either disappearing and being replaced by woodland or else are becoming more and more like lowland grasslands characterized by an impoverished, trivial flora.

References

Autonome Provinz Bozen-Südtirol (2010) Statistisches Jahrbuch für Südtirol. Landesinstitut für Statistik – ASTAT, Bolzano

Celesti-Grapow L, Alessandrini A, Arrigoni PV, Banfi E, Bernardo L, Bovio M, Brundu G, Cagiotti MR, Camarda I, Carli E, Conti F, Fascetti S, Galasso G, Gubellini L, La Valva V, Lucchese F, Marchiori S, Mazzola P, Peccenini S, Poldini L, Pretto F, Prosser F, Siniscalco C, Villani MC, Viegi L, Wilhalm T, Blasi C (2009) Inventory of the non-native flora of Italy. Plant Biosystems 143(2):386–430

Conti F, Manzi A, Pedrotti F (1992) Libro Rosso delle piante d'Italia. Associazione Italiana per il World Wildlife Fund, Roma

Dalla Torre KW, Sarnthein L (1906-1913) Die Farn- und Blütenpflanzen von Tirol, Vorarlberg und Liechtenstein, 4 Teile. Wagner'sche Universitäts-Buchhandlung, Innsbruck

Hausmann F (1851-1854) Flora von Tirol. Wagner, Innsbruck

Kottek M, Grieser J, Beck C, Rudolf B, Rubel F (2006) World map of the Köppen-Geiger climate classification updated. Meteorologische Zeitschrift 15(3):259–263

Niklfeld H (1971) Bericht über die Kartierung der Flora Mitteleuropas. Taxon 20:545–571

Niklfeld H (2003) Für die Flora Südtirols neue Gefäßpflanzen (1): Ergebnisse der floristischen Kartierung, vornehmlich aus den Jahren 1970-1998. Gredleriana 2(2002):271–294

Prosser F (2001) Lista Rossa della flora del Trentino. Museo Civico di Rovereto

Pyšek P, Richardson DM, Rejmánek M, Webster GL, Williamson M, Kirschner J (2004) Alien plants in checklists and floras: towards better communication between taxonomists and ecologists. Taxon 53(1):131–143

Schneeweiß GM, Niklfled H (in preparation) Arealtypen der Gefäßpflanzen Österreichs

Wilhalm T (2011) Ergänzungen und Korrekturen zum Katalog der Gefäßpflanzen (4). Gredleriana 11:71–82

Wilhalm T (2012) Die Erforschung der Flora Südtirols im 20. Jahrhundert. In: Huber O et al (eds) Die Botanik in Südtirol und angrenzenden Gebieten im 20. Jahrhundert: Eine bibliographische Rundschau. Die Gärten von Schloss Trauttmansdorff (Hrsg.). Edition Raetia, Bozen, pp 83–96

Wilhalm T, Hilpold A (2006) Rote Liste der gefährdeten Gefäßpflanzen Südtirols. Gredleriana 6: 115–198

Wilhalm T, Stockner W, Tratter W (2003) Für die Flora Südtirols neue Gefäßpflanzen (2): Ergebnisse der floristischen Kartierung, vornehmlich aus den Jahren 1998-2002. Gredleriana 2(2002):295–318

Wilhalm T, Zemmer F, Beck R, Stockner W, Tratter W (2005) Für die Flora Südtirols neue Gefäßpflanzen (3): Ergebnisse der floristischen Kartierung, vornehmlich aus den Jahren 2002-2004. Gredleriana 4(2004):381–412

Wilhalm T, Niklfeld H, Gutermann W (2006) Katalog der Gefäßpflanzen Südtirols. Veröffentlichungen des Naturmuseums Südtirol, 3. Folio, Wien-Bozen.

Wilhalm T, Beck R, Schneider-Fürchau E, Tratter W (2009) Ergänzungen und Korrekturen zum Katalog der Gefäßpflanzen Südtirols (3). Gredleriana 9:105–118

Floristic Biodiversity in Trentino

2

Alessio Bertolli, Francesco Festi, and Filippo Prosser

Abstract

The Flora biodiversity of wild vascular plants in Trentino (=Trento province, North Italy) is briefly presented on the basis of the data (over 1,125,000 records) collected and stored by Museo Civico of Rovereto since 1990. A relevant number of "quadrants" (ca. 35 km^2) or 1 × 1 km grid maps are shown. About 2400 wild species are present in Trentino, including naturalized (ca. 200) and not confirmed/extinct ones (ca. 50). Further about 600 species are recorded as casual. Floristic richness distribution is uneven, ranging from 450 species per quadrant in high siliceous mountains to 1300 in limestone in Southern Adige Valley. 150 species are endemic in the Alps, and among these 50 are steno-endemic of the Prealps between Lombardy and Veneto. *Callianthemum kernerianum, Gentiana brentae, Primula recubariensis, Nigritella buschmanniae, Erysimum aurantiacum* are very local species, and the last two grow only in Trentino. About 1/3 of the flora of Trentino is threatened and it is listed in the Trentino's Red List of vascular plants. In comparison with historic data, 50 species are not confirmed and probably extinct. Red List species are mainly concentrated at lower altitude, where human impact is higher. There are about 200 alien naturalized species, including some 100 species that are native in other Italian regions. Nearly 50 come from Americas and over 20 from Asia. Chorological and life form spectra are given as well as some elaboration using Ellenberg's ecological indicators.

Keywords

Flora biodiversity • Vascular plants • Trentino • Alps • Chorological spectra

A. Bertolli • F. Festi • F. Prosser (✉)
Fondazione Museo Civico di Rovereeto, Largo S. Caterina 41, 38068 Rovereto, Italy
e-mail: prosserfilippo@fondazionemcr.it

F. Pedrotti (ed.), *Climate Gradients and Biodiversity in Mountains of Italy*, Geobotany Studies, https://doi.org/10.1007/978-3-319-67967-9_2

2.1 Introduction

Trentino is completely included in the southern part of the Alps (Fig. 2.1). It occupies an area of 6207 km^2 (2% of whole Italy). The territory is completely mountainous and mainly steep, apart from the principal valley floors and some uplands. Trentino reaches neither the watershed nor the southern margin of the Alps. In spite of this it hosts a wide landscape diversity. The geological diversity is high, since both acid (igneous and metamorphic) and basic (limestone, dolomite) rocks are equally widespread (Fig. 2.2). Calcareous schist lacks. The height ranges from 3766 m a.s.l. of Mt Cevedale to 66 m a.s.l. of Lake Garda. The southern part of Trentino is wetter and has a little thermic range (suboceanic climate); the northern one is dryer and has a higher thermical range (subcontinental climate) (Figs. 2.3 and 2.4). The species that better locally represent this gradient are *Fagus sylvatica* L., mainly distributed in the South of Trentino, and *Pinus cembra* L., mainly distributed in the North (Figs. 2.5 and 2.6). A detailed phytoclimatic classification and the related map of the territory are given by Gafta and Pedrotti (1998). Vegetation belts from submediterranean to nival are present. Several local phytosociological vegetation maps at different scale are available from Trentino: we can mention for instance the map of Trento and surrounding by Pedrotti (1981).

Trentino has about 525,000 inhabitants (0.9% of the population of whole Italy); so the density is 85 inhabitants/km^2 (the one of whole Italy is 198 inhabitants/km^2). The distribution of population is uneven: 50% of population is located below 400 m a.s.l. Minghetti (1999) gives a map of the naturalness of vegetation of Trentino.

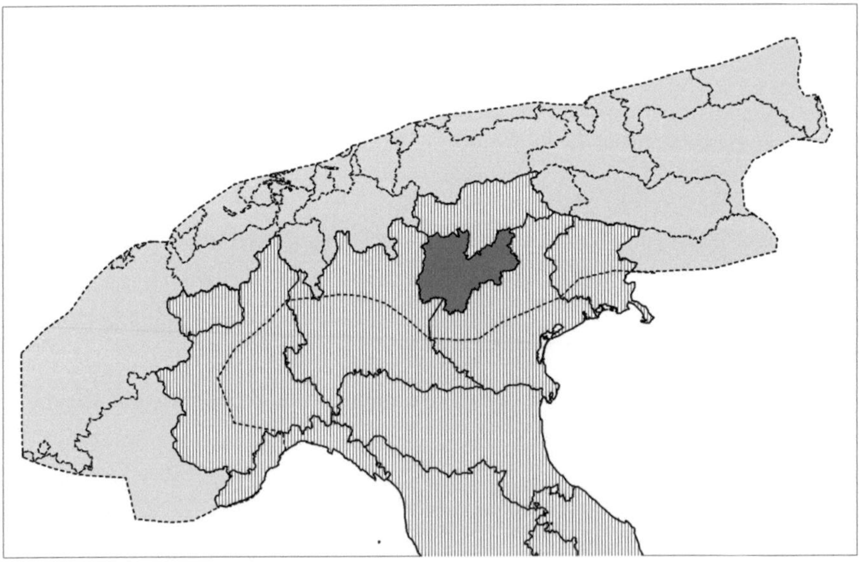

Fig. 2.1 Map of the Alps (*dot line*) and Northern Italy (*vertical lines*) with location of Trentino (*grey*)

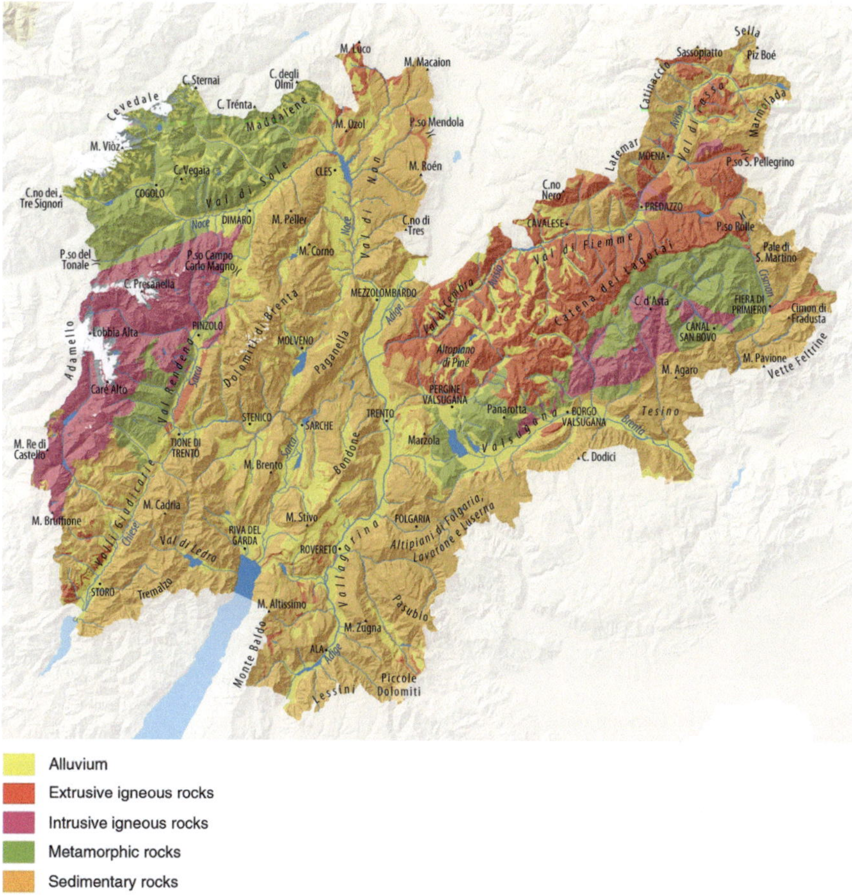

Fig. 2.2 Schematic geological map of Trentino

The first explorers of the flora of Trentino were Pier Andrea Mattioli (1501–1578) and Francesco Calzolari (1522–1609) in the sixteenth century. However, modern scientific exploration began in Trentino only in the nineteenth century thanks to Francesco Facchini (1788–1852) and to Enrico Gelmi (1855–1901). The main historic work concerning the flora of Trentino is "Flora von Tirol" by Dalla Torre and Sarnthein (1900–1913), in which all previous floristic information was gathered and critically analysed. The "Flora of Tyrol" was the basis of later important works such as "La nostra Flora" by Giuseppe Dalla Fior (1969) and the deep phytogeographic analysis by Vittorio Marchesoni (1946–1947). Franco Pedrotti carried out botanic studies in Trentino over 60 years. Thanks to his activity, we have today a wide knowledge of the vegetation of this territory.

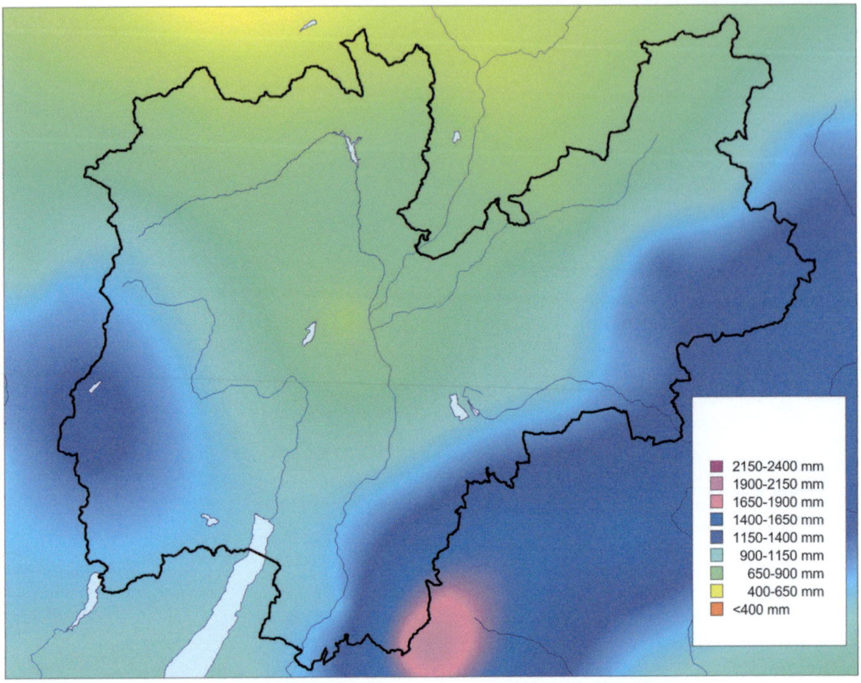

Fig. 2.3 Rainfall in Trentino (mm/year)

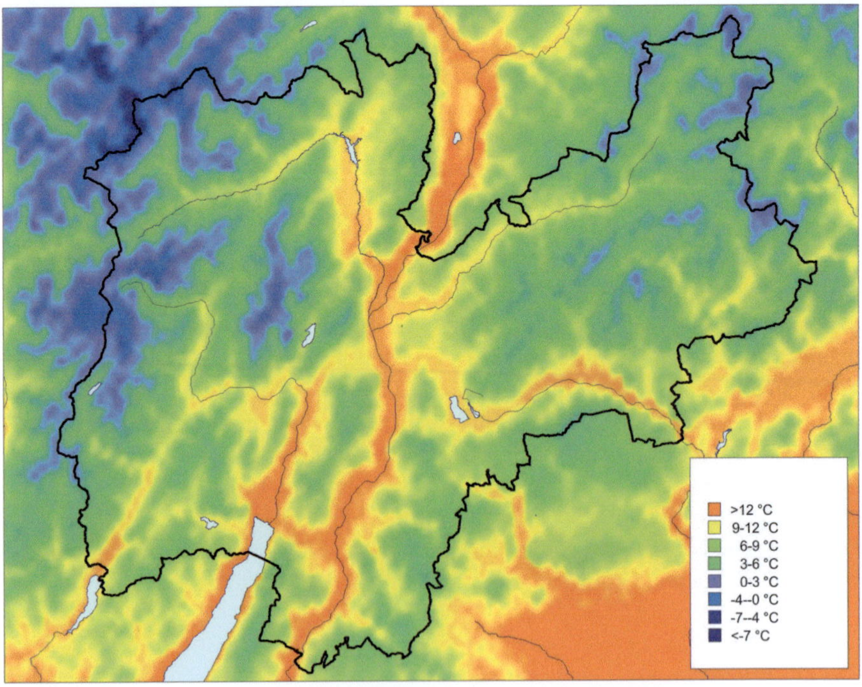

Fig. 2.4 Average Annual Temperature in Trentino (°C)

Fig. 2.5 *Fagus sylvatica* wood in Southern Trentino (Mt Pasubio), and grid map of this species in Trentino

Fig. 2.6 *Pinus cembra* wood in Northern Trentino (Fassa Valley), and grid map of this species in Trentino

2.2 Monitoring the Flora of Trentino

This project by Museo Civico of Rovereto, started in 1990, aimed to map all vascular plants growing in Trentino. At present, we are going on collecting data with the purpose of completing and updating the dataset. Data sampling is partly carried out by volunteers, partly funded by public Institutions (e.g. Servizio Protezione Natura of Provincia Autonoma of Trento, Natural Parks, etc.), rarely by private Institutions (Banks, etc.). Data of all wild species (about 2400 indigenous or naturalized and over 600 casual) are collected all over the area. All data are computerized and most of them are geocoded (Table 2.1). A dedicated computer program (Festi 2005) is being used to manage the data which, at present, are not available on the web.

Table 2.1 Stored data in the plant databases of Museo Civico of Rovereto: number of records on August 2013

Field data from observation lists[a]	846,366
Field data from single observation[a]	75,687
Data from Herbarium ROV[a]	36,221
Data from other Herbaria[b]	25,846
Data from literature[b]	141,144
TOTAL	1,125,264

[a]Geocoded data
[b]Referred to quadrant

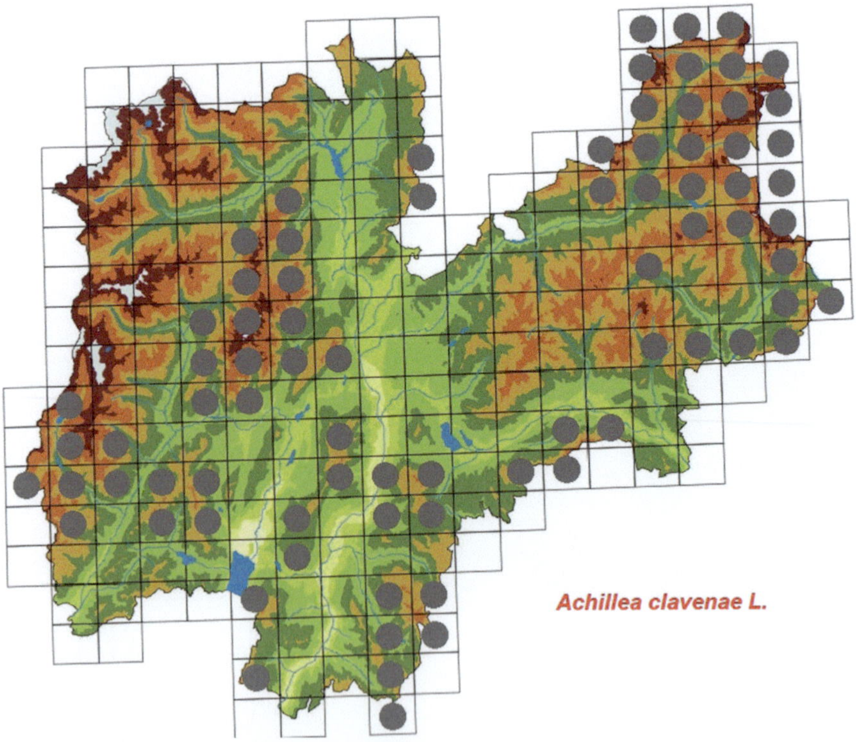

Fig. 2.7 Example of distribution grid map ("quadrants") of *Achillea clavenae* in Trentino

Data can be seen and for example the growing range of species according to recent data in "quadrant" can be shown (after Ehrendorfer and Hamann 1965) (Fig. 2.7). It is also possible to view both historic and recent data: in this case, the map shows the reduction of the growing area of the species (Fig. 2.8). Published atlases following this grid are the ones of Paneveggio-Pale di S. Martino Natural Park (Festi and Prosser 2000) and Adamello-Brenta Natural Park (Festi and Prosser 2008).

Most data are linked to geographic objects (generally segments or areas, rarely points) more detailed than quadrants using a GIS program. The geographic objects can be reduced to their centroids or to a grid cell (generally an 1 × 1 km cell) containing the centroid. This process is not completely rigorous since some data could be shifted. Furthermore, the data were collected with the aim of obtaining complete lists of species for each quadrant, and not for each cell: it means that inside each quadrant the collection of data was substantially casual. Even so, the obtained point maps and 1 × 1 km grid maps are far more detailed than the quadrant maps. The following example of point map shows the spreading of the alien species *Geranium sibiricum* L. in the last two decades. Detailed point maps are provided in two atlases: Atlas of Flora of Mt Baldo (Prosser et al. 2009: see Fig. 2.9) and Atlas of Flora of Mt Pasubio (Prosser et al. 2012: see Fig. 2.10).

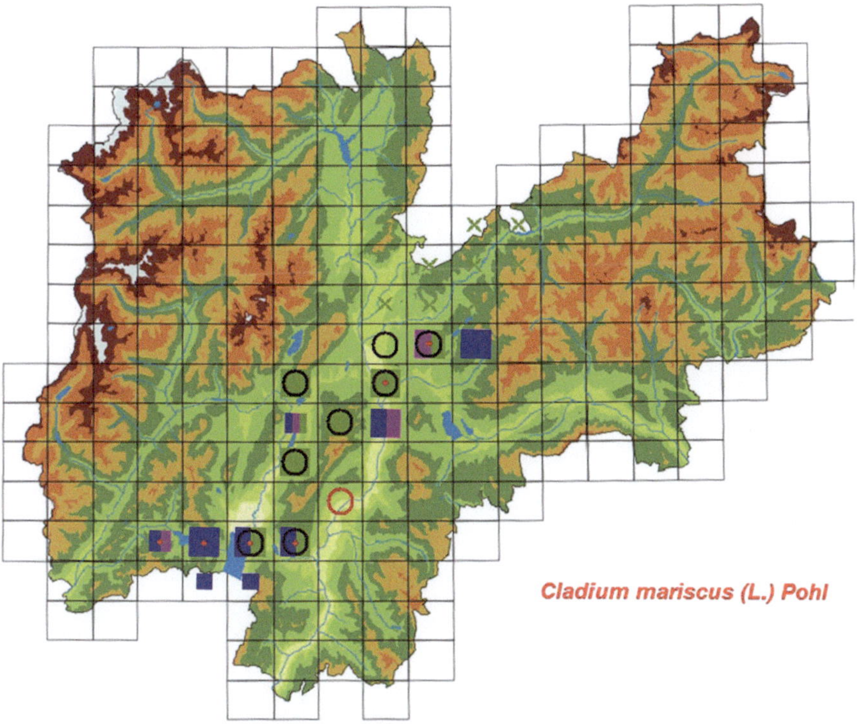

Fig. 2.8 Example of distribution grid map ("quadrants") of *Cladium mariscus* with both historic (*squares*) and recent (*circles*) data on map grid ("quadrant") in Trentino: *Cladium mariscus*

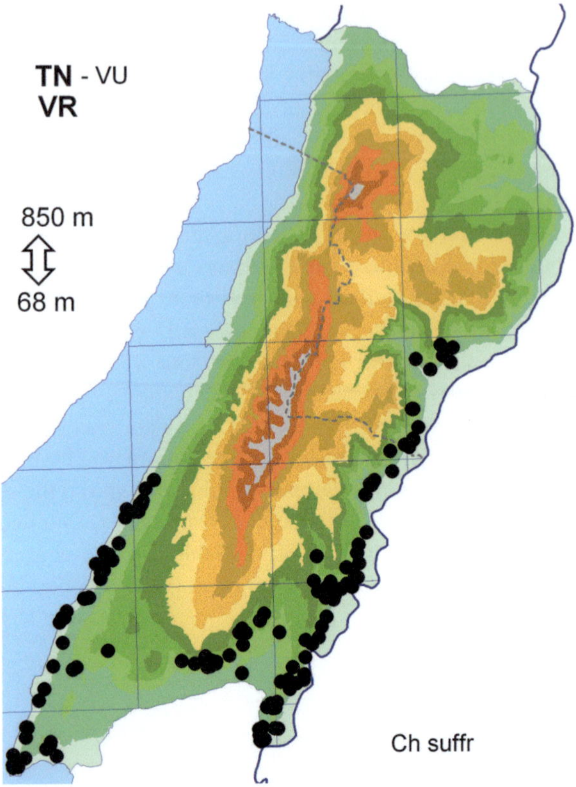

Fig. 2.9 Example of detailed point map (after Prosser et al. 2009): *Satureja montana*

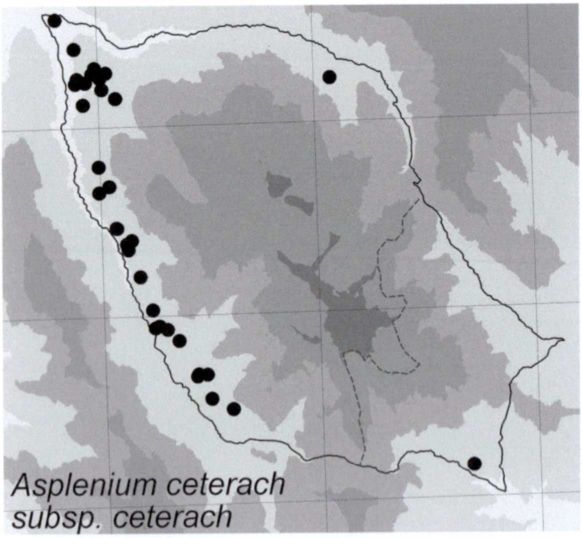

Fig. 2.10 Example of detailed point map (after Prosser et al. 2012): *Ceterach officinarum*

The altitude range of each recorded species can also be displayed; it is useful to compare the vertical distribution of species (Fig. 2.11).

2.3 Floristic Richness of Trentino

At present we can estimate the number of vascular plants (species and subspecies) growing as native or naturalized in Trentino: nearly 2400 taxa. In comparison with 1800s, about 50 of these taxa are missing, whereas about 200 have entered in recent times and got naturalized. In addition, over 600 taxa are recorded only as casual. The total number of taxa reported from Trentino (excluding the erroneous ones) is about 3000 (Fig. 2.12).

It is interesting to observe the number of species (including subspecies) recorded in each quadrant of Trentino: see Fig. 2.13. Unconfirmed ancient data are not considered here. The part of Trentino with the highest plant biodiversity is the middle-southern one, and in particular the Adige Valley from Trento southwards (Fig. 2.14), the southern side of Brenta Group, the Sarca Valley, with some further scattered quadrants: the Chiese Valley, the Sole Valley and the Primiero Valley. On the other hand, the lowest plant biodiversity can be observed in the siliceous chains of Lagorai, Adamello-Presanella and Ortles-Cevedale (Fig. 2.15). Although this result could be basically right, we notice that the areas with the highest number of

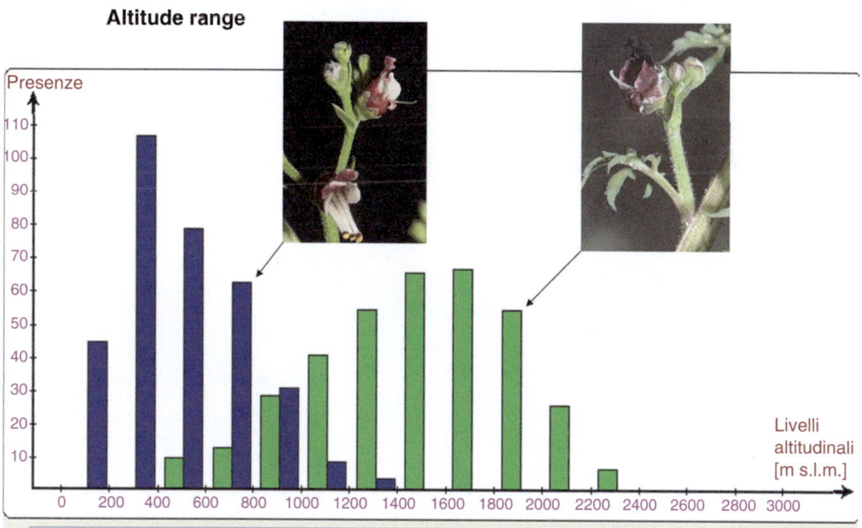

Specie	Media	Dev. st.	Freq. max.	A metri	Tot. segn.	Perc. segn./esc.	Q min.	Q max.
Scrophularia juratensis	1530	415	67	1700	370	12.1	450	2430
Scrophularia canina	510	270	107	300	338	11.4	74	1400

Fig. 2.11 Comparison of altitude range of the altitude vicariant species *Scrophularia canina* and *S. juratensis*

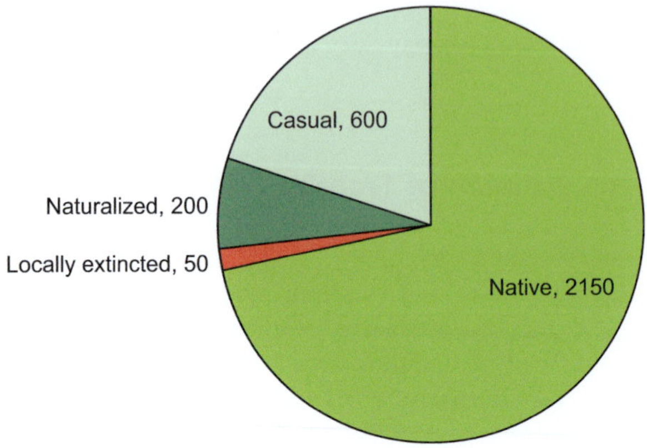

Fig. 2.12 Number of species reported from Trentino divided in main groups

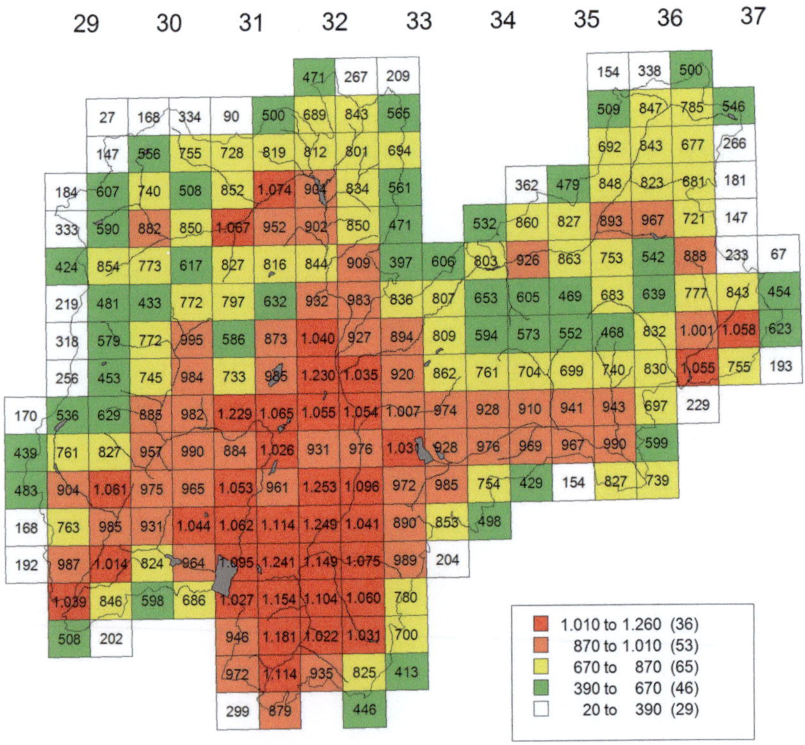

Fig. 2.13 Number of taxa (species and subspecies) recorded in each quadrant of Trentino

species match well with the ones with the highest field exploration: see Fig. 2.16. In this image we display the ratio between the total number of recorded data and total number of taxa for each quadrant: it corresponds to the number of times each taxon

Fig. 2.14 Example of landscape with high phytodiversity: Adige Valley. Some traits are the wide elevation range starting from valley plain with warm habitats, limestone rocks and the presence of habitats disturbed by man

Fig. 2.15 Example of landscape with low phytodiversity: Adamello-Presanella Group. Some traits are siliceous rocks, the exclusion of warm habitats, the absence of human impact

was on average recorded in each quadrant. This ratio shows the level of field exploration of the territory.

2.4 Chorological Spectrum of the Flora of Trentino

We compare the chorological spectrum of the Flora of Trentino (mainly after Pignatti 1982) and of Italy as a whole (Pignatti 1994). From the graph (Fig. 2.17) we infer that Boreal, Eurasiatic and Orophyte S-European species are more represented in Trentino than in the whole of Italy, whereas Endemic, Mediterranean Mountain, Steno-Mediterranean species are less represented.

We show in detail some maps of chorological groups. The first one concerns the Mediterranean s.l. species (after Pignatti 1982) growing in Trentino (Fig. 2.18). The map shows clearly the thermophilous areas and corridors in Trentino: Alto Garda and the Sarca Valley up to Tione (and the Rendena Valley with lower values), the Adige Valley (also lateral Valleys Ronchi, Vallarsa and Terragnolo are clearly

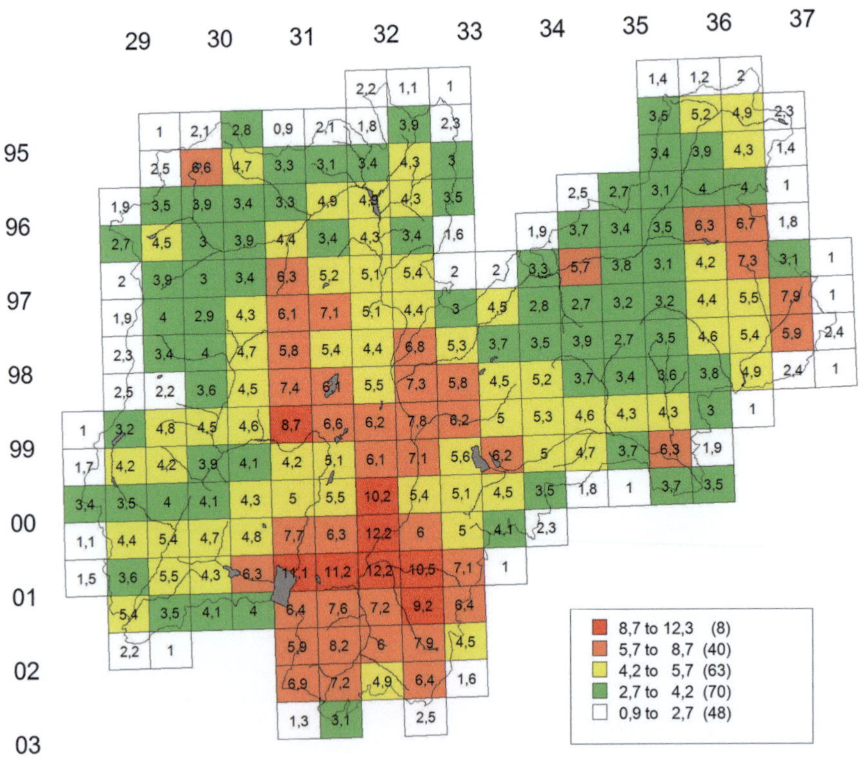

Fig. 2.16 Ratio between total number of data recorded and total number of taxa in each quadrant. This ratio shows the field exploration of the territory

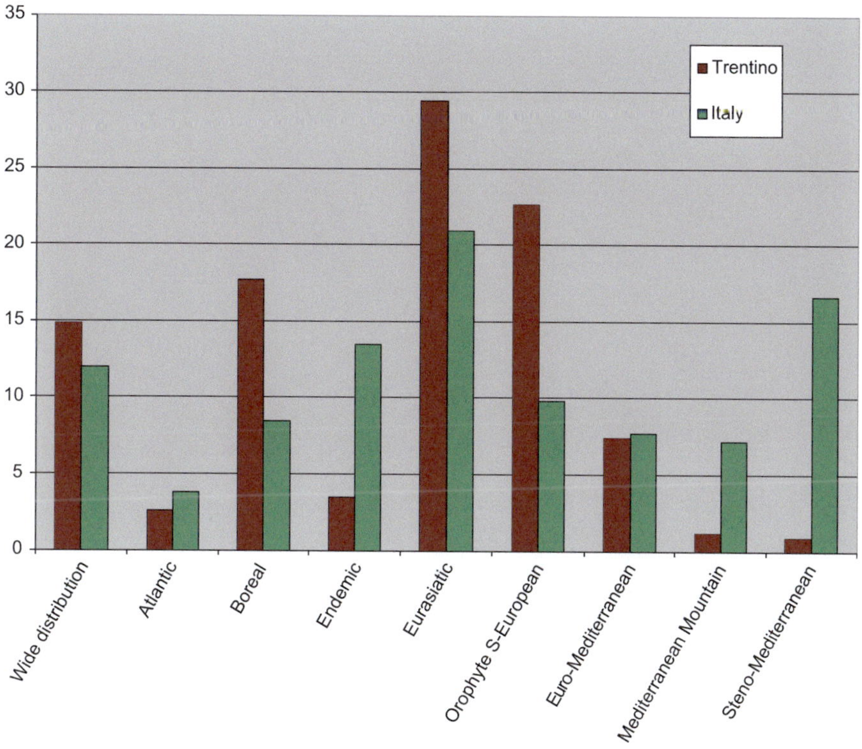

Fig. 2.17 Comparison of chorological spectrum of the Flora of Trentino and of whole Italy. Values in percentage

viewed), the Chiese Valley, Valsugana, Primiero, Cembra, Fiemme, Non and Sole Valleys.

In the second example, among the Mediterranean s.l. only the Steno-Mediterranean species are considered (Fig. 2.19). The map shows the areas with submediterranen influence, that is the surroundings of Lake Garda and further north along the lower Sarca Valley up to Lake Terlago. Along the Adige Valley the occurrence of Steno-Mediterranean species is rare and limited to some areas, such as Avio, Serravalle, Mori (notice here the connection with Lake Garda area through Lake Loppio), Rovereto, Pomarolo: scattered occurrences reach further North the boundary with Alto Adige in the surroundings of Mezzocorona. As typical Steno-Mediterranean species we can mention *Phillyrea latifolia* L. (Fig. 2.20), growing in Trentino only in the limited areas above mentioned.

A further example concerns the Artic-Alpine species (Fig. 2.21). This group was included in the Boreal species in Fig. 2.19. The map shows clearly the mountainous areas of Trentino and it is somewhat the opposite of the map of Mediterranean s.l. species (Fig. 2.24). As typical example of Artic-Alpine species we can mention

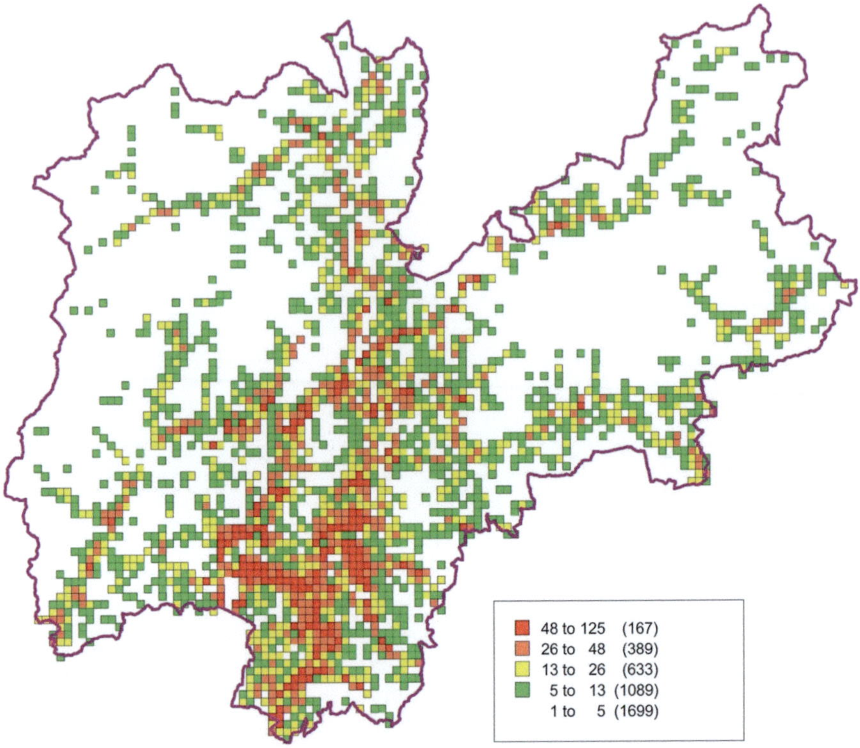

Fig. 2.18 1 × 1 km grid map showing the distribution of the Mediterranean s.l. species in Trentino

Carex norvegica Retz., growing in Trentino only in a few areas of Ortles-Cevedale Group in upper Rabbi and Pejo Valleys (Fig. 2.22).

2.5 Endemic Species in Trentino

Due to their importance, we focus our attention on endemic species, even though—as already pointed out—the percentage of endemic species is lower in Trentino than in Italy as a whole. In Trentino some 150 species among the about 500 endemic species of the Alps reported by Aeschimann et al. (2004) can be found. Among them 50 grow in a little part of the Pre-Alps between Lombardy and Veneto. This small group of steno-endemic species, as shown in Fig. 2.23, has an uneven distribution in Trentino: most of them are localized in the southwestern part of the territory (Ledro Alps), at the boundary with the Prealps of Brescia (Lombardy). Further limited concentrations of steno-endemic species are located in the southern part of Brenta Group and in the Piccole Dolomiti. Some parts of Trentino lack completely steno-endemic species: this is the case with the Cembra Valley and with the northern part of the Non Valley, both

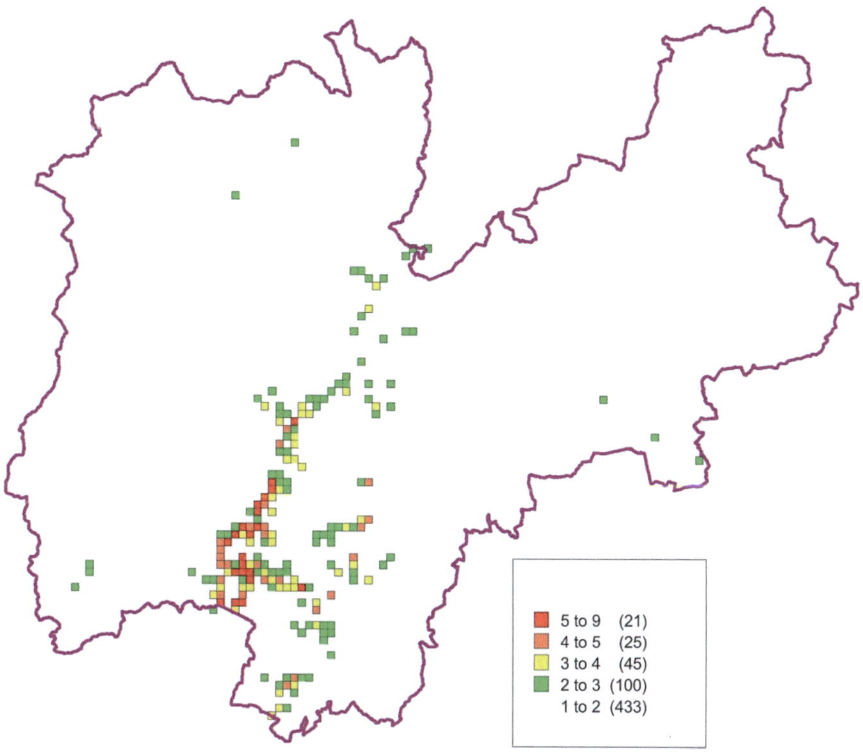

Fig. 2.19 1 × 1 km grid map showing the distribution of the Steno-Mediterranean species in Trentino

on siliceous rocks. The uneven distribution of endemic species is generally explained by the quaternary glaciation: on the southern side of the Alps some peaks were not covered with ice and here species of the previous age either survived or evolved into new very local species under isolation.

With the exception of the mainly apomictic *Alchemilla, Hieracium, Ranunculus auricomus* group, *Rubus* and *Taraxacum*, only two species are endemic of Trentino: *Nigritella buschmanniae* Teppner & Ster and *Erysimum aurantiacum* (Leyb.) Leyb. The survey of the flora of Trentino offered some unexpected outcomes as far as the endemic species are concerned. We found and described three new species: *Festuca austrodolomitica* (Pils and Prosser 1995), *Primula recubariensis* (Prosser and Scortegagna 1998) and *Gentiana brentae* (Prosser and Bertolli 2008) (Fig. 2.24). The last species grows nearly only in the Brenta Group, but it was found also on Mt Bondolo at the boundary with Lombardy. Only in 2013 the wonderful steno-endemic *Pinguicula poldinii*, already known in Friuli-Venezia Giulia and Veneto, was found in Trentino (Bertolli and Prosser 2013).

Figure 2.25 shows the 1 × 1 km grid map of the ca. 150 endemic species of the Alps (after Aeschimann et al. 2004) occurring in Trentino. This map matches quite

Fig. 2.20 *Phillyrea latifolia*, tipical Steno-Mediterranean species

well with the quadrant map of stenoendemic species (Fig. 2.20). Besides the Ledro Alps, other mountains of southern Trentino show a high concentration of endemic species of the Alps: southern Adamello, Mt Baldo, Piccole Dolomiti, Pasubio, Stivo-Bondone Chain, Vette di Feltre; also the high values in the Brenta Group, Pale di S. Martino and Dolomiti di Fassa are outstanding.

2.6 Alien Flora of Trentino

As already shown, nearly 200 species have been naturalized in Trentino over the last two centuries. About 100 belong to the alien flora of Italy (Celesti-Grapow et al. 2010); among them nearly 50 come from Americas and over 20 from Asia. Over the last decades about 100 species have immigrated to Trentino from other Italian territories where they are considered native. In Fig. 2.26 the 1 × 1 km grid map of the naturalized species in Trentino is presented. This map is similar to the map of the Mediterranean s.l. species (Fig. 2.24) and it clearly shows the areas with the largest human impact. All the main valleys of Trentino are affected by a considerable amount of naturalized alien species while the mountainous areas are nearly free from this invasion. As a typical example of naturalized species we mention *Rumex kerneri* Borbás (Fig. 2.27), that probably started spreading from the central part of Trentino to the rest of northern Italy about 30 years ago.

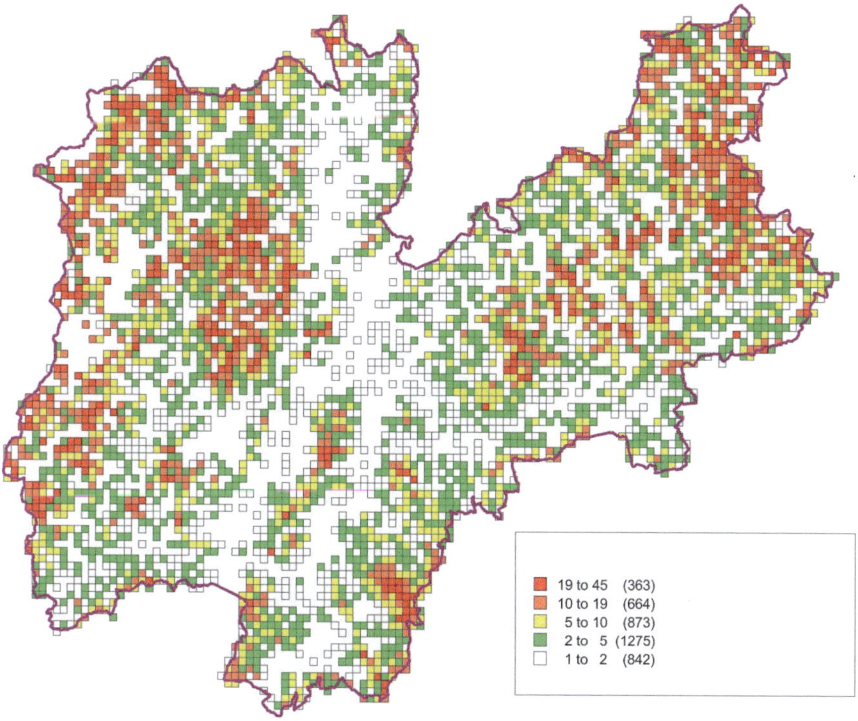

Fig. 2.21 1 × 1 km grid map showing the distribution of the Artic-Alpine species in Trentino

Fig. 2.22 *Carex norvegica*, typical example of Artic-Alpine

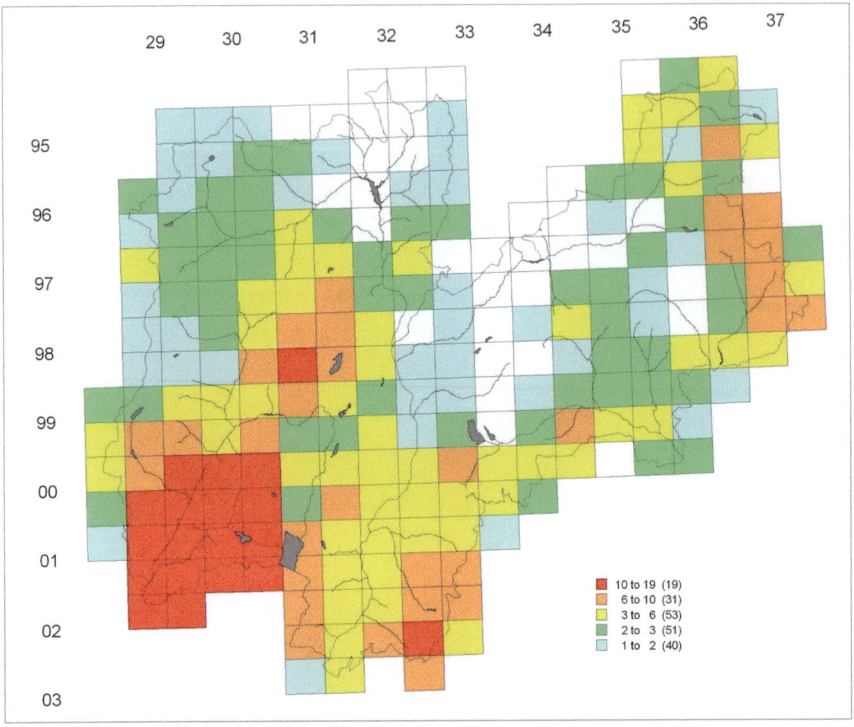

Fig. 2.23 Distribution of steno-endemic species in Trentino

Fig. 2.24 *Gentiana brentae* Prosser & Bertolli, is an endemic species of Trentino. It grows nearly only in Brenta Group

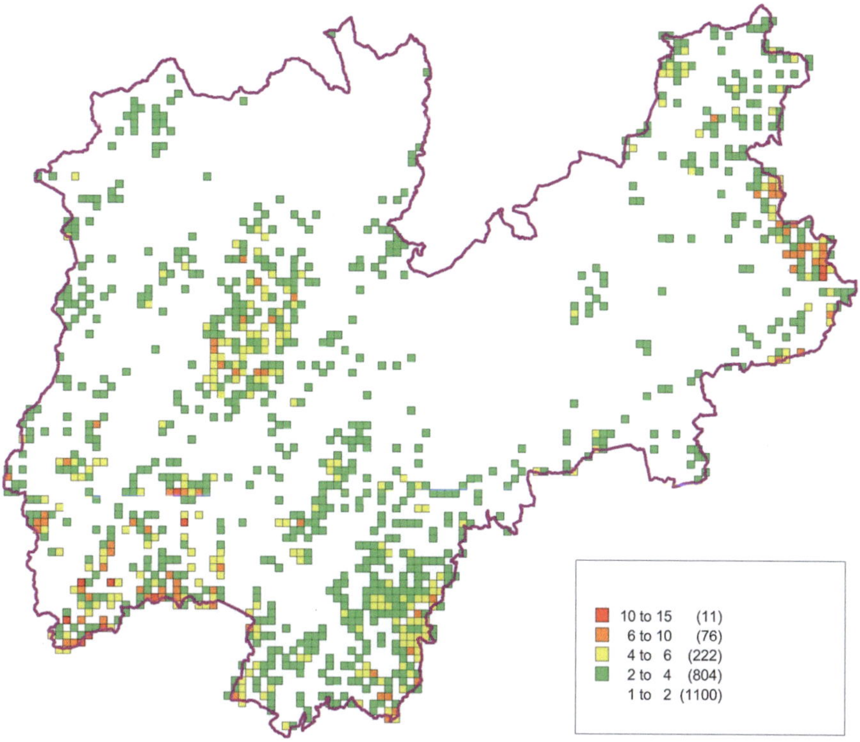

Fig. 2.25 1 × 1 km grid map showing the distribution of the endemic species of the Alps (after Aeschimann et al. 2004) occurring in Trentino

2.7 Red List of the Vascular Flora of Trentino

Prosser (2001) provided the province of Trento with its own Red List of vascular plants. In the following years the Red List was continuously updated. The present Red List shows that nearly 1/3 of Trentino's Flora is threatened at different degrees and 3% (ca. 50 species) is extinct in this province. Figure 2.28 shows the species number of Red List of Trentino for each quadrant. It is clear that the lowlands are the most affected by the presence of threatened species: this part of territory suffers from significant landscape changes and here human impact is more concentrated. Dry meadows of the hills and swampy places of the valley plains have nearly disappeared and their flora is mainly in danger.

We can display the same data converting the I.U.C.N. categories as follows: NT = 1, VU = 2, EN = 3, CR = 4, DD = 1 and referring them to 1 × 1 cells. We obtain a map (Fig. 2.29) showing in more details the location of threatened species (and their habitats) in Trentino. The following areas can be identified: dry areas

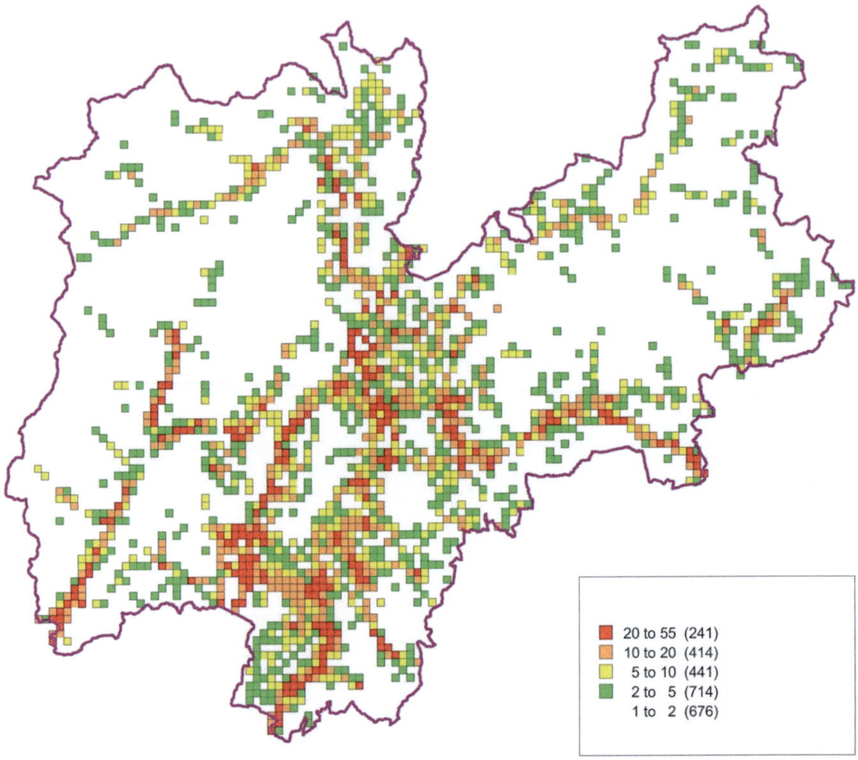

Fig. 2.26 1 × 1 km grid map showing the distribution of the alien species of Trentino

near Arco, Riva, Rovereto, Pomarolo and scattered also in other parts of Trentino; moist areas of Mt Sous (upper Non Valley), Madonna di Campiglio, the Cembra Valley, the Duron Valley, Paneveggio-Bellamonte, Tesino, Pinè, Mezzocorona. Not all these areas are at present protected.

Figure 2.30 shows the number of extinct species for each quadrant in Trentino. The data are obviously based on literature and ancient herbaria. Cities such as Trento and Rovereto are characterized by a high loss of species. We can explain this with their expansion in the last decades, but also with the availability of far more historic data from this area than from other territories. As a matter of fact, many ancient botanists lived in Trento and Rovereto, and they collected data mainly in the neighbourhood of their residences. An example of extinct species in Trentino is represented by *Orchis laxiflora* Lam. (Fig. 2.31), a typical species of wet meadows at a low altitude.

Fig. 2.27 *Rumex kerneri* Borbás (a, b), example of alien species naturalized in Trentino

2.8 Ecological Indicators After Ellenberg et al. (1991)

We present also some quadrant maps concerning Ellenberg's indicators (Ellenberg et al. 1991). In particular we consider indicators of pH (Fig. 2.32), temperature (Fig. 2.33) and humidity (Fig. 2.34). The map of pH indicator is substantially similar to the geologic map (Fig. 2.2). The map of temperature indicator is substantially similar to the map of Mediterranean s.l. species (Fig. 2.20). The map of humidity indicator is somewhat the opposite of the map of pH indicator: quadrants on siliceous underground show clearly higher values than quadrants on sedimentary underground.

2.9 Life Form Spectrum of the Flora of Trentino

In Fig. 2.35 we compare the life form spectrum of the Flora of Trentino (after Pignatti 1982) with the life form spectrum of the Flora of Italy as a whole (Pignatti 1994). Major differences can be seen in Hemicryptophytes, widespread in alpine

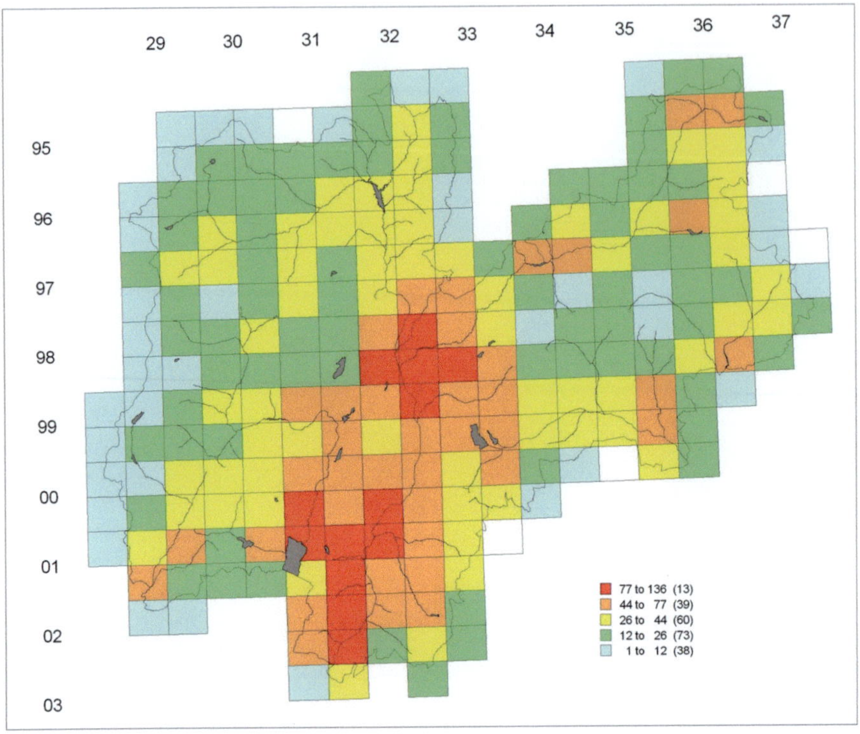

Fig. 2.28 Red List Species of Trentino for each quadrant (after Prosser 2001 and unpublished update)

grasslands and more represented in Trentino, and in Therophytes, widespread in dry habitats of Mediterranean areas and more represented in Italy.

We present also the 1x1 km grid map of Phanerophytes of Trentino (Fig. 2.36), showing a remarkable richness of tree species along the valleys at a lower altitude, and the similar map of Therophytes (Fig. 2.37), showing a concentration of annual species in the lower areas of Trentino.

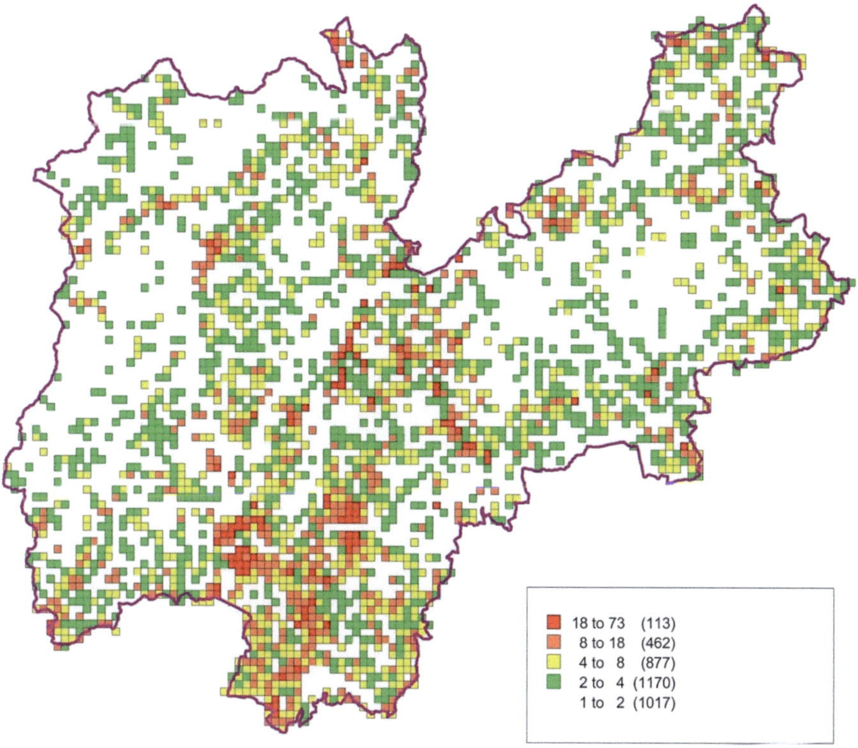

Fig. 2.29 Map of Red List Index in Trentino. The categories of the species are converted as following: NT = 1, VU = 2, EN = 3, CR = 4, DD = 1

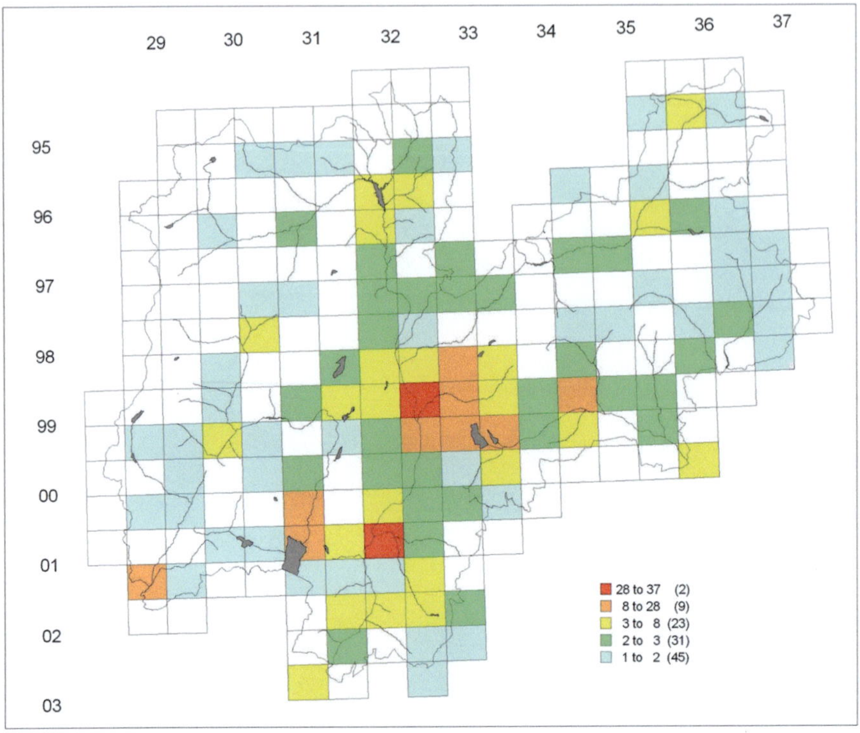

Fig. 2.30 Number of extinct species for each quadrand in Trentino

Fig. 2.31 Example of extinct species in Trentino: *Orchis laxiflora*, specimen from Riva del Garda (1896, leg. Enrico Gelmi, TR)

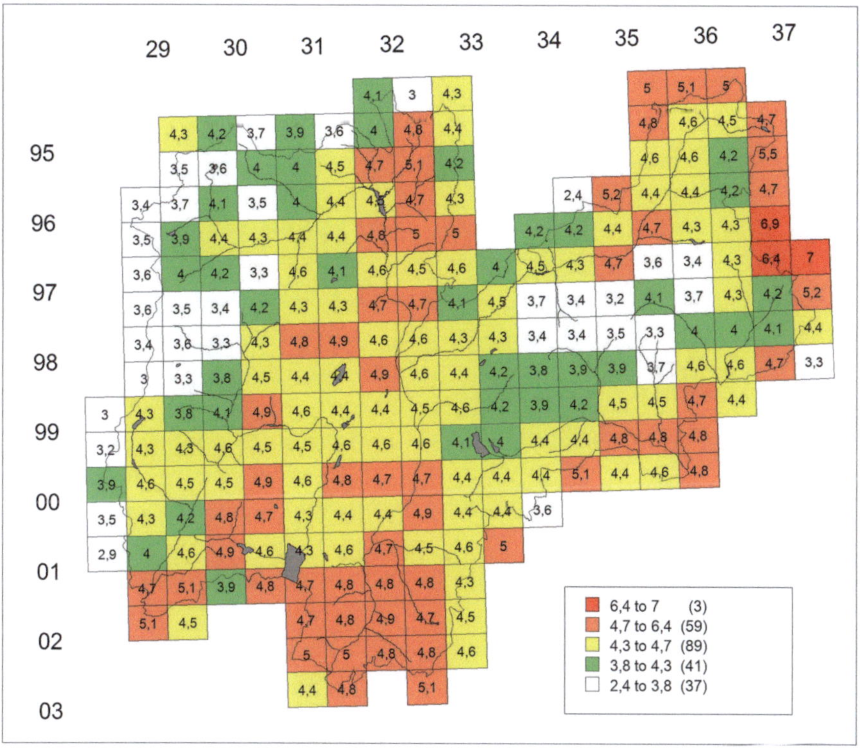

Fig. 2.32 Quadrant map of Trentino of Ellenberg's pH indicator

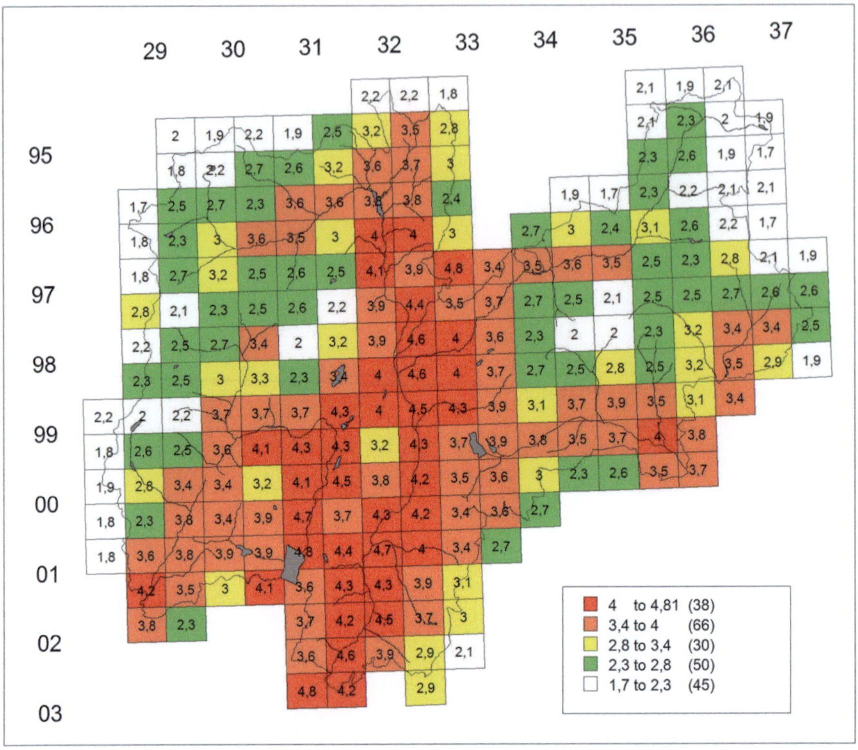

Fig. 2.33 Quadrant map of Trentino of Ellenberg's temperature indicator

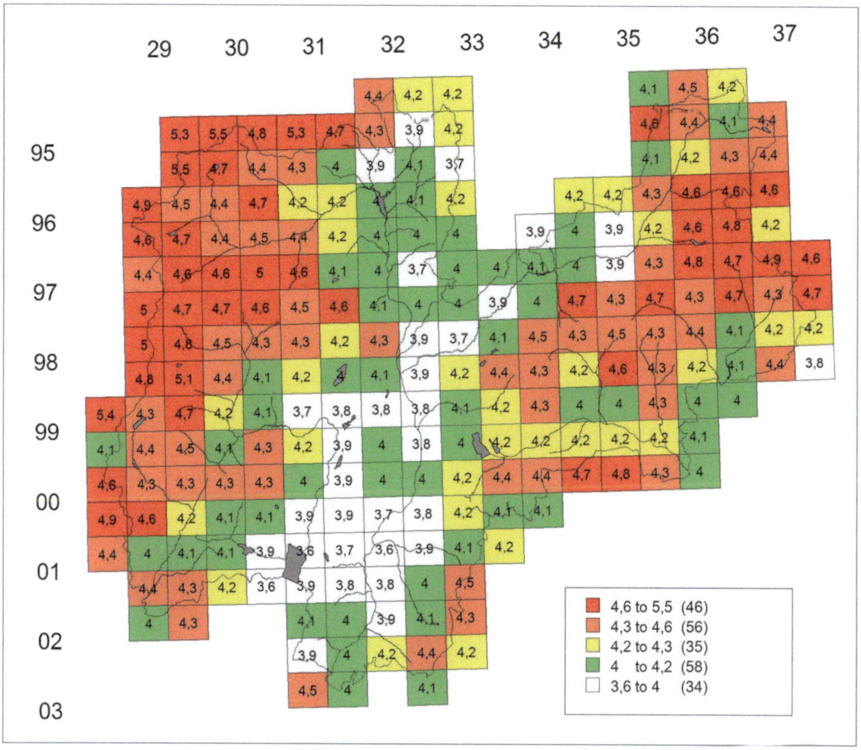

Fig. 2.34 Quadrant map of Trentino of Ellenberg's humidity indicator

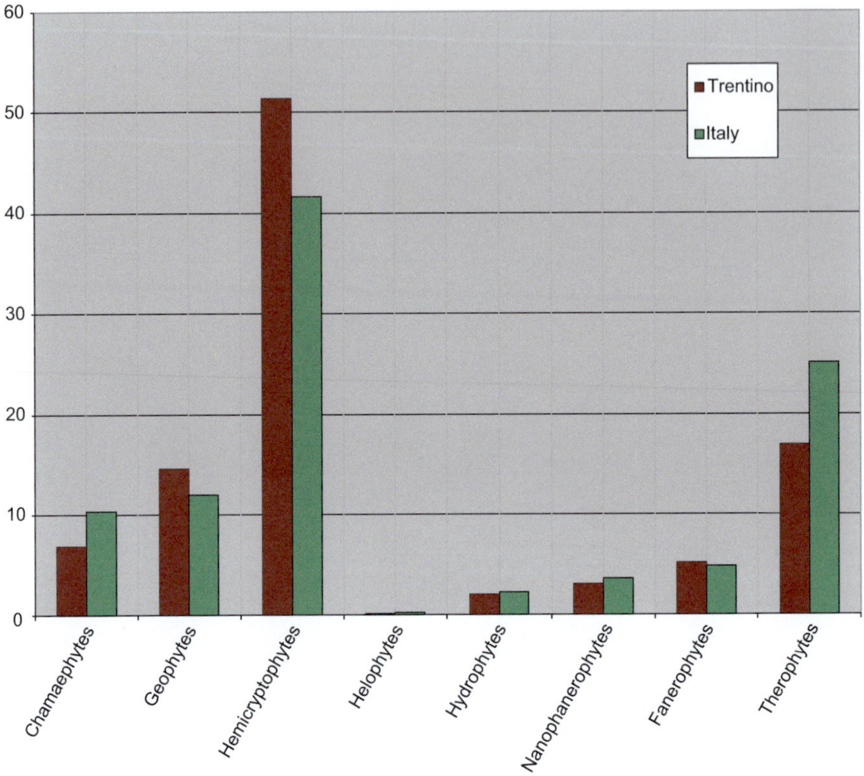

Fig. 2.35 Comparison of life form spectrum of the Flora of Trentino and of whole Italy. Values in percentage

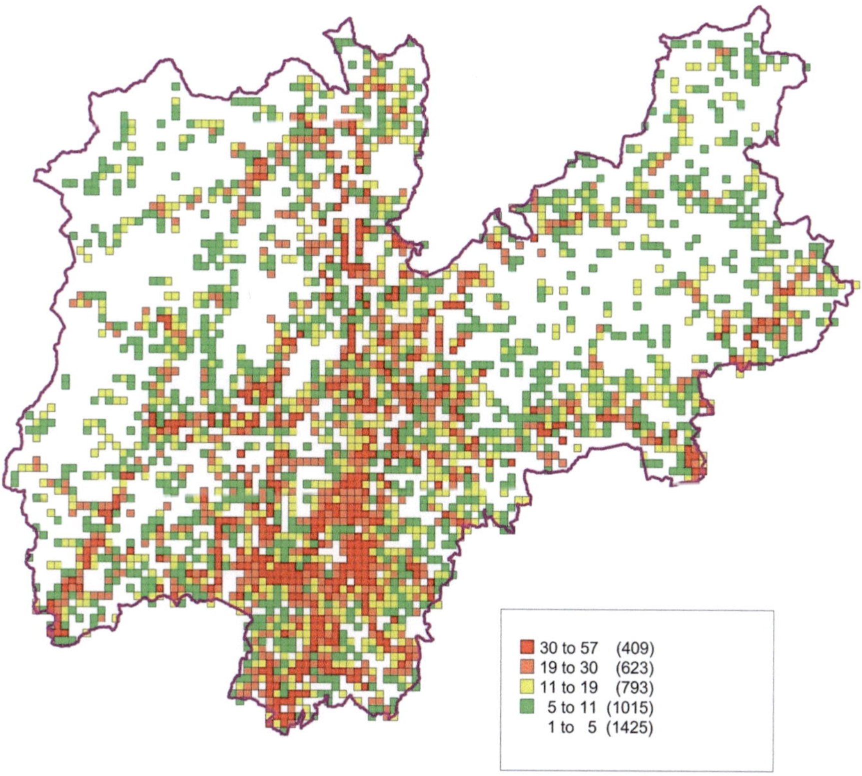

Fig. 2.36 1 × 1 km grid map of Phanerophytes of Trentino

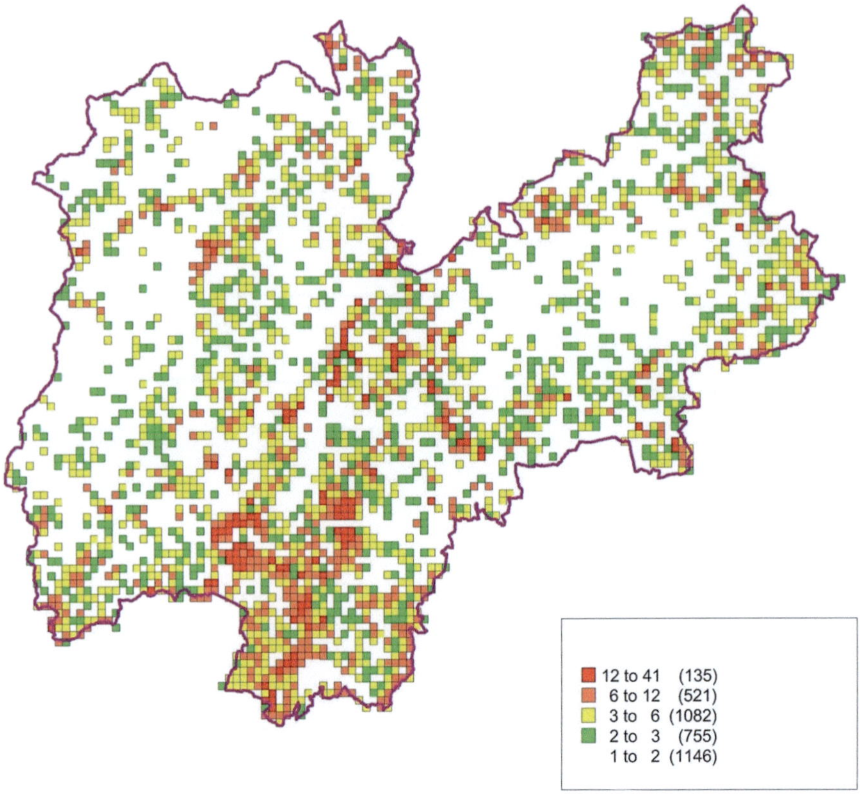

Fig. 2.37 1 × 1 km grid map of Therophytes of Trentino

References

Aeschimann D, Lauber K, Moser DM, Theurillat J-P (2004) Flora Alpina. 3 voll. Haupt, Bern/ Zanichelli, Bologna

Bertolli A, Prosser F (2013) Notula 2008. Pinguicula poldinii Steiger & Casper (Lentibulariaceae), specie nuova per il Trentino-Alto Adige. Inf Bot Ital 45(2):302–303

Celesti-Grapow L, Pretto F, Carli E, Blasi C (eds) (2010) Flora vascolare alloctona e invasiva delle regioni d'Italia. Casa Editrice Università La Sapienza, Roma, 208 pp.

Dalla Fior G (1969) La nostra flora. Ed. Monauni, Trento

Dalla Torre KW, Sarnthein L (1900–1913) Flora der Gefürsteten Grafschaft Tirol, des Landes Vorarlberg und der Fürstenthumes Liechtenstein. Vol. I. Die Litteratur der Flora (1900); Vol. IV. Die Farn- und Blütenpflanzen (Pteridophyta et Siphonogama) – 1 (1906);2 (1909);3 (1912);4 (1913). Verlag der Wagner'schen Universitäts-Buchhandlung, Innsbruck

Ehrendorfer F, Hamann U (1965) Vorschläge zu einer floristischen Kartierung von Mitteleuropa. Ber Deutsch Bot Ges 78:35–50

Ellenberg H, Weber HE, Düll R, Wirth V, Werner W, Paulissen D (1991) Zeigerwerte von Pflanzen in Mitteleuropa. Scripta Geobotanica 18:1–248

Festi F (2005) CFT, un programma di gestione della Cartografia Floristica Tridentina, versione 2004. Ann Mus civ Rovereto 20(2004):107–126
Festi F, Prosser F (2000) La Flora del Parco Naturale Paneveggio-Pale di S. Martino. Atlante corologico e repertorio delle stazioni. Suppl Ann Mus civ Rovereto, Sez St, Arch, Sc nat, 13 (1997) 438 pp
Festi F, Prosser F (2008) Flora del Parco Naturale Adamello-Brenta. Documenti del Parco nr. 17, ed. Osiride, Rovereto, 606 pp
Gafta D, Pedrotti F (1998) Fitoclima del Trentino-Alto Adige. Studi Trent Sc Nat, Acta Biol 73 (1996):55–111
Marchesoni V (1946–1947) Analisi fitogeografica degli elementi floristici del bacino atesino. Mem Museo Trid Sci Nat, 7 (1946):4–78; 8 (1947):3–40
Minghetti P (1999) Naturalness of vegetation of Trentino-Alto Adige (with map 1:250,000). Centro di Ecologia Alpina, report 20:1–35
Pedrotti F (1981) Carta della vegetazione del Foglio Trento. Consiglio Nazionale delle Ricerche, Roma. AQ/1/17
Pignatti S (1982) Flora d'Italia. 3 Voll., Edagricole, Bologna
Pignatti S (1994) Ecologia del Paesaggio. UTET, Torino, 225 p
Pils G, Prosser F (1995) Festuca austrodolomitica spec. nova, a new species of the F. halleri group (Poaceae) from the SE. Alps Plant Syst Evol 195:187–197
Prosser F (2001) Lista Rossa della Flora del Trentino. Pteridofite e Fanerogame. LXXXIX pubblicazione del Museo Civico di Rovereto, 107 p
Prosser F, Bertolli A (2008) A new species of Gentiana sect. Calathianae (Gentianaceae) from the Brenta Group, European Alps, Italy. Willdenowia 38:423–431
Prosser F, Scortegagna S (1998) Primula recubariensis, a new species of Primula sect. Auricula Duby endemic to the SE Prealps, Italy. Willdenowia 28:27–45
Prosser F, Bertolli A, Festi F (2009) Flora illustrata del Monte Baldo. Ed. Osiride, Rovereto, 1240 p
Prosser F, Bertolli A, Casarotto N, Festi F, Scortegagna S, Zara F (2012) Atlante della flora vascolare del Pasubio (province di Trento e Vicenza). Ann Mus Civ Rovereto 27 (2011):87–220

Vegetation Series Along Climatic Gradients in the Central Southern Alps (Trentino- Alto Adige Region)

3

Franco Pedrotti

Abstract

Here is presented the distribution of the vegetation series along the climatic gradients of the sector of the central Alps, southern side. That corresponds to the Trentino-Alto Adige Region. They were distinct zonal, intrazonal, azonal and extrazonal series. The vegetation of the Trentino-Alto Adige Region is caracterized by 128 vegetation series. This high number is the consequence of multiple causes, such as latitudinal extension, lithological substrates, geomorphological conditions and phytoclimate. It demonstrates the complexity and the great phytogeographic diversity of the territory under study, with a very marked gradient, passing from the prealpic to the endoalpic sector.

Keywords

Vegetation series • Zonal • Intrazonal • Azonal and extrazonal series • Climatic gradients • Trentino-Alto Adige Region • Alps

The goal of this contribution is to examine the distribution of the vegetation series along the climatic gradients of the sector of the central Alps, southern side, that corresponds to the Trentino-Alto Adige Region. The area comprises the catchment basin of the Adige River and some lesser basins nearby, namely the Brenta, Bacchiglione, Sarca and Chiese.

F. Pedrotti (✉)
University of Camerino, Camerino, Italy
e-mail: franco.pedrotti@unicam.it

F. Pedrotti (ed.), *Climate Gradients and Biodiversity in Mountains of Italy*,
Geobotany Studies, https://doi.org/10.1007/978-3-319-67967-9_3

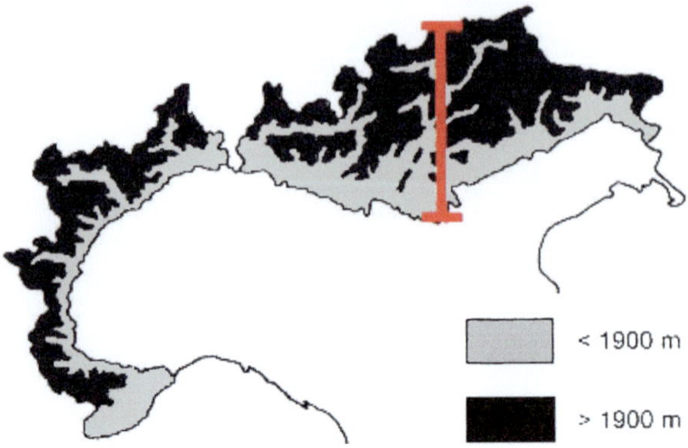

Fig. 3.1 Map of the Italian Alps; in correspondence with the basin of the Adige (Trentino—Alto Adige Region) from the Brenner Pass to the Padanian plain, the alpine chain extends for 200 km, which is the maximum dimension reached

3.1 Physical and Geographic Characteristics

This territory is typically mountainous, with groups and chains interrupted by very deeply cut valleys. The lowest altitude is in the Padania Plain (40–50 m s.l.m.); at Riva del Garda the altitude is 67 m, while the highest is the Ortles peak (3905 m). The zone thus delimited extends between a northern latitude of 45°40′ and 47°06′ and between an eastern longitude of 10°23′ and 12°28′. From north to south, in correspondence with the line from the Brenner Pass to Verona, the Alps extend for about 200 km (Fig. 3.1), which is the maximum dimension that the chain of the Alps reaches in Italy, both in terms of the western Alps (Piemonte and Val d'Aosta) and the eastern ones (Veneto and Friuli-Venezia Giulia), where it is always much narrower.

The orographic conformation and the consequent disposition of the mountain massifs condition the layout of the valleys, which are oriented according to parallels and meridians and according to intermediate positions. Valleys oriented according to parallels are Val Venosta, Val di Sole, Val di Genova, Valle del Sarca between Tione and Sarche, and Valsugana; these valleys show very marked ecological dissymmetry between the slope with southern exposure, which is sunnier and warmer, and the one with northern exposure, which is colder and more humid. The side with southern exposure is characterized by forests of *Quercus petraea* (*Luzulo niveae-Quercetum petraeae*) and steppe-like meadows of the *Stipo-Poion xerophilae* alliance. Val Venosta is also characterized by *Astragalo-Pinetum sylvestris* pinewoods; instead the side with northern exposure is occupied by forests of *Vaccinio-Piceion*. This phenomenon happens only in the valleys situated more in the interior of the alpine chain (Val Venosta and Val di Sole) and not in the more external ones, such as the Valle del Sarca between Tione and Sarche and the Valsugana, with deciduous forests, where the difference between the two sides only involves lower altitudes on the slopes with northern exposure, and higher altitudes on those with southern exposure.

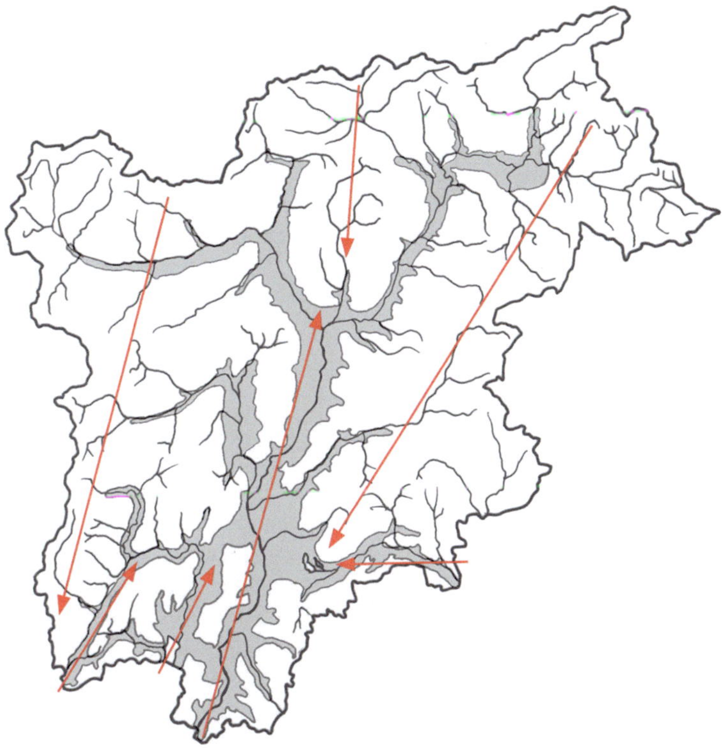

Fig. 3.2 Map of the Trentino-Alto Adige Region; the Valle dell'Adige and the lateral valleys permit a number of deciduous species to penetrate from south and east toward the interior of the alpine chain, among them *Quercus pubescens*, *Ostrya carpinifolia*, and *Fraxinus ornus* (area indicated in *grey*). The mountain massifs favor the southward descent of continental species. The *arrows* indicate the directions from south and east to north, and from north to south

The valleys oriented according to meridians, such the Val d'Adige, Valle dell'Isarco, Val Sarentina, Val di Non, Val Rendena, Valle del Chiese, and Valle del Sarca between Toblino and Riva del Garda, have sides with similar characteristics. They have the same type of vegetation: in Val d'Adige and Valle dell'Isarco there is *Fraxino orni-Ostryetum carpinifoliae,* in Valle del Sarca between Toblino and Riva del Garda, there is not only *Fraxino orni-Ostryetum carpinifoliae* but also *Celtidi australis-Quercetum ilicis*. In Val Rendena there are the *Galio laevigati-Carpinetum betuli* and *Salvio glutinosae-Fraxinetum excelsioris* associations. Similarly, in the valleys with intermediate layout, like the Cembra, Fiemme, Fassa and Pusteria valleys, there are no significant differences between the two sides.

In addition, the distribution of the vegetation in the region is also strongly conditioned by the mountainous massifs and the main valleys (Fig. 3.2). The incised valleys (Val d'Adige) favour the penetration toward the north, that is, into the innermost parts of the alpine chain, of the associations of deciduous trees, both from the hilly plane (*Fraxino orni-Ostryetum*, *Luzulo-Quercetum petraeae*, ecc.) and from the montane one (*Carici albae-Fagetum sylvaticae*, *Cardamino-Fagetum sylvaticae*, ecc.). Instead, the mountain groups over 2000 m of the southern part of the region (Adamello and Lagorai), through the effect of their mass, cause a southward descent of the conifer associations of the montane and submontane belts (the *Piceion excelsae* and the *Rhododendro-Vaccinion* associations) of herbaceous species of the alpine belt (*Caricion curvulae*). This fact is very clear on the maps of the distribution of the vegetation series reported below (Figs. 3.7a, 3.8, 3.9 and 3.10a).

3.2 Lithology

In this territory there is an elevated number of lithological substrates, which can be grouped into four summary categories, on the basis of potential trophicity (content of bases) of the overlying soils (Pedrotti and Gafta 2003): (1) limestones and dolomites unit (limestones, dolomites, marbles, chalks, marls, Werfen formations) (Prealps and Dolomites); (2) metamorphic and magmatic intrusive unit (granites, diorites, granodiorites, granitoids, tonalite, gneiss, paragneiss, orthogneiss, phyllites, micaschits, chlorite-schists, basalts) (Ortles, Cevedale, Adamello and others); (3) magmatic effusive unit and related slaking materials (rhyolites, dacites, rhyodacites, ignimbrites, porphyroids, andesites, Val Gardena sandstones) (Lagorai); (4) glacial deposits (morains), fluvial deposits (alluvial terraces) and peaty deposits (mires), that occupy valley bottoms, high plains and parts of the slopes.

3.3 Climate

This entire zone belongs to the Eurosiberian region with temperate macrobioclimate (Rivas Martìnez 1996). The bioclimatic study conducted by Gafta and Pedrotti (1998) proposes the distinction of seven thermic types (lower and upper mesotemperate, lower and upper supratemperate, orotemperate, cryo-orotemperate, and glacial), four ombric types (xeric, subxeric, subhumid and humid) and three hygric-pluvial continentality types (prealpic, alpic and inner-alpic). While the thermotypes correspond to the succession of the bioclimatic belts (colline, montane, subalpine, alpine and glacial), the types of continentality mentioned above are identified with two sectors, termed suboceanic and subcontinental.

3.4 Vegetation Series

Vegetation series identification and description was conducted with reference to knowledge about the real vegetation of the Trentino-Alto Adige Region (Pedrotti 1965–1968; Pedrotti et al. 1974; Schiechtl and Stern 1976, 1982; Mayer 1977;

Pitschmann et al. 1980; Pedrotti 1981, 1982, 1992, 1993; Peer 1995; Prosser 1997; Prosser and Sarzo 2003; Pedrotti et al. 2003, etc.) and in reference to the summary lithological map, the phytoclimatic map, the map of the potential vegetation, and the map of the megageoseries and the series of the Trentino-Alto Adige Region (Gafta and Pedrotti 1998; Pedrotti and Gafta 2003; Pedrotti 2013). In some locations specific studies were conducted on the vegetation series and their cartography (Lago di Loppio, Gafta and Pedrotti 1994; Torbiera di Fiavé, Pedrotti 1997; Laghestel di Piné, Pedrotti 2004).

The series of vegetation were grouped on the basis of the phytogeographic criterion of zonality, that is, in the following categories: zonal, extrazonal, intrazonal and azonal (Ivan 1979). The attribution of each series to one of the categories mentioned was evaluated on the basis of the distribution (chorology) and synecology of the plant formations in which it is found. The distinction between intrazonal and azonal vegetation employed the followed criterion as well: the former is present in only one bioclimatic belt, while the latter usually is spread through more than one belt.

The series heads of the zonal and extrazonal vegetation form a climatic climax (on the metamorphic and magmatic-intrusive substrates) or an edaphic-climatic climax (on substrates that are calcareous-dolomitic and magmatic-effusive). This interpretation arises from the concept of "placor" (Ivan 1979), which means the climax vegetation that forms in intermediate site conditions (flatland habitats on substrates that are neither carbonatic, nor very acidic, nor influenced by the phreatic surface or by water stagnation, etc.) in a given bioclimatic belt.

The vegetation series are named after the dominant species, and the sigmetum after the association that heads the series, as explained in detail in Pedrotti (in print). The definition of a series (sigmetum) was done following the proposal of Rivas-Martìnez (1987). The descriptor refers to bioclimatic belt, chorography, ombrotype, edaphic affinities, and the most evolved communities of the series (associations at the head of the series). After the descriptor, other observations have been added: the lithological typology of the substratum (calcareous, dolomitic, magmatic, or silicatic), the sector of the Trentino-Alto Adige Region (pre-alpic, alpic, inner alpic) where the series is present, and indications on the locations where the series with limited distribution are found in the Trentino-Alto Adige Region. The term "alpic" is used in reference to the distribution of a series in the alpine chain; it is also used to indicate one subdivision into three sectors of the territory of the Trentino-Alto Adige Region on the basis of the index of hygric continentality: alpic, inner-alpic, and prealpic.

3.4.1 Climatic Gradients and Vegetation Series

The analysis of the phytoclimate of the Trentino-Alto Adige Region was conducted by Gafta and Pedrotti (1998), including preparation of a phytoclimatic map on a scale of 1: 250,000. For the analysis of the climatic gradients, the following aspects

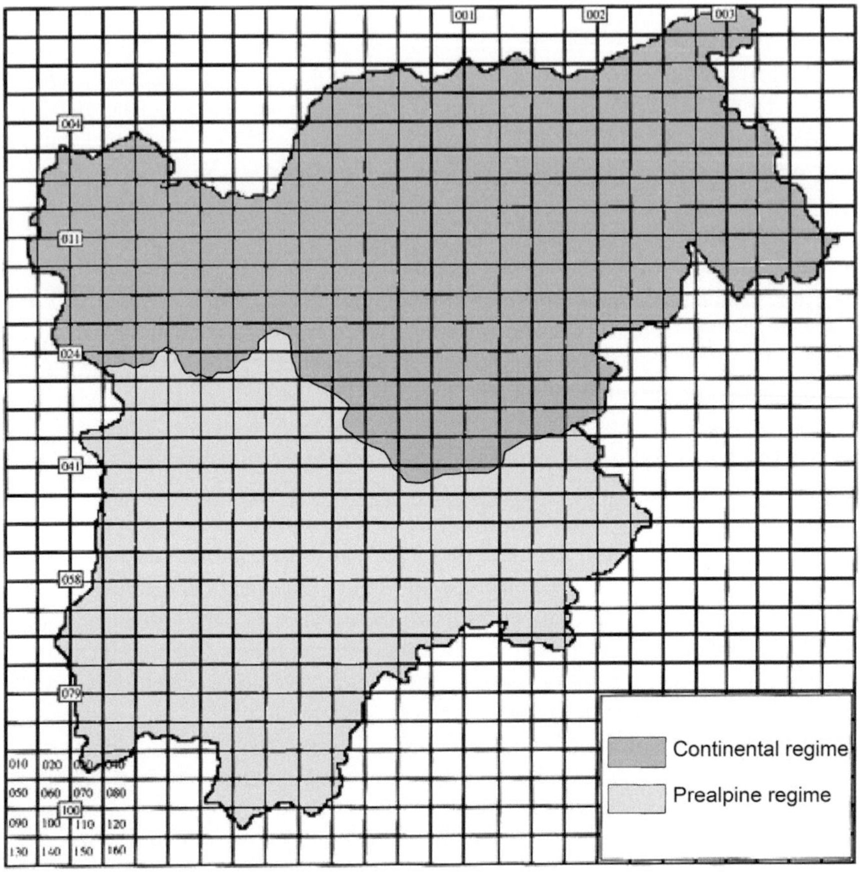

Fig. 3.3 Map of the annual pluviometric regime types in the Trentino-Alto Adige Region. The zone with continental regime is characterized by summer unimodal and summer bimodal regimes. The one with prealpine regime is marked by autumn bimodal and spring bimodal regimes. The map refers to the UTM network (from Gafta and Pedrotti 1998, modified)

were taken into consideration here: annual pluviometric regime, ombrotype, thermal continentality and pluvio-hygric continentality.

Regarding the annual pluviometric regime, the Trentino-Alto Adige Region can be attributed to two distinctly different zones, the first with a continental regime and the second with a padano-prealpic one; the line of separation divides the region into two almost equal parts, with the continental regime to the north, and the padano-prealpic one to the south (Fig. 3.3).

There are four ombrotypes: xeric, subxeric, subhumid and humid. The first two are limited to the Valle Venosta, which constitutes a "climatic island" in the region (Fig. 3.4). The remainder of the region has a humid ombrotype, except a few valley bottoms, such as the Val d'Adige, Valle dell'Isarco, Val di Non, Val di Sole,

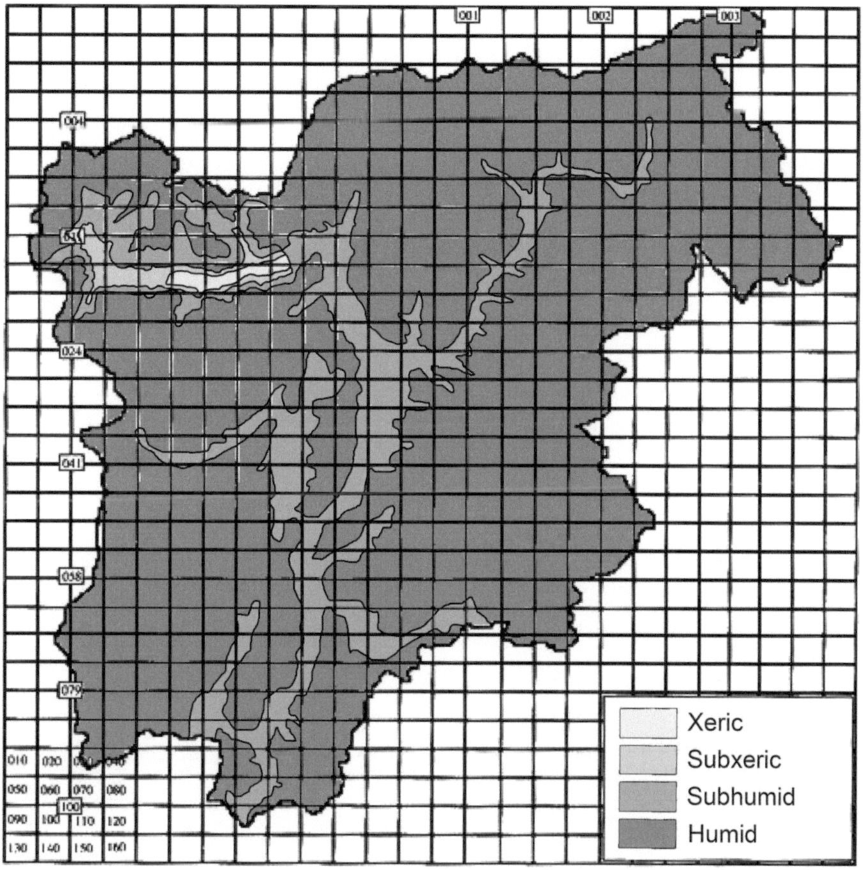

Fig. 3.4 Map of the ombric types of Trentino-Alto Adige Region; the map is referred to the UTM network (from Gafta and Pedrotti 1998, modified)

Valsugana and Valle del Sarca from Toblino to Garda, which have a subhumid ombrotype (Fig. 3.4).

There are two types of thermic continentality, the suboceanic one throughout the region, and the subcontinental one in some main valleys (Fig. 3.5).

The index of pluvio-hygric continentality are attributed to three categories, prealpic, alpic and inner-alpic. On the basis of this index, the area of the southern slope of the Alps can be divided into three sectors (Fig. 3.6): (1) prealpic, which corresponds to southern Trentino, occupying the entire Valsugana and some secondary valleys, the Val d'Adige as far as a bit beyond Salorno, the entire Val di Non and part of the Val di Sole, the Valli del Sarca (excluding the high Val di Genova) and that of the Chiese (excluding the Val di Fumo); (2) alpic, north of the previous sector as far as the watershed line; (3) inner-alpic, itself divided into two zones, a western one (the Val Venosta) and an eastern one that is more articulated (the high

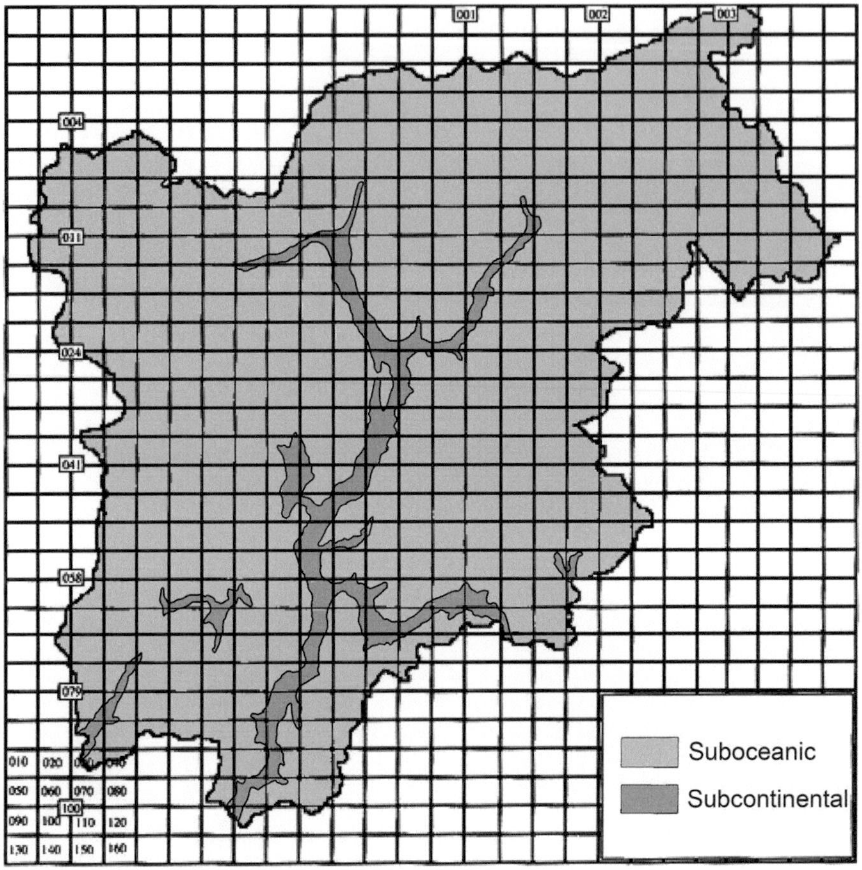

Fig. 3.5 Map of the thermic continentality types from Trentino-Alto Adige Region; the map is referred to the UTM network (from Gafta and Pedrotti 1998, modified)

Val Pusteria, the Dolomites, mountain ridges between Bressanone and Val Sarentina). This division is more detailed than that of previous authors, including Mayer (1977) and Ozenda (1985, 2013), who identified only two zones, alpic and prealpic.

To examine the climatic gradients and related vegetation series, it is possible to consider the entire region, or megageoseries. The first case is a "general" gradient from north to south, through the entire region, which comprises the three sectors—prealpic, alpic, and inner-alpic—each of which is formed by certain vegetation series. The second case refers to megageoseries. The 128 vegetation series listed previously can be united into megageoseries, units that group the complete altitudinal sequence of the zonal and extrazonal vegetation series in a given sector and district (Theurillat 1992). In the Trentino-Alto Adige Region, it is possible to identity 10 megageoseries and 19 submegageoseries, which are

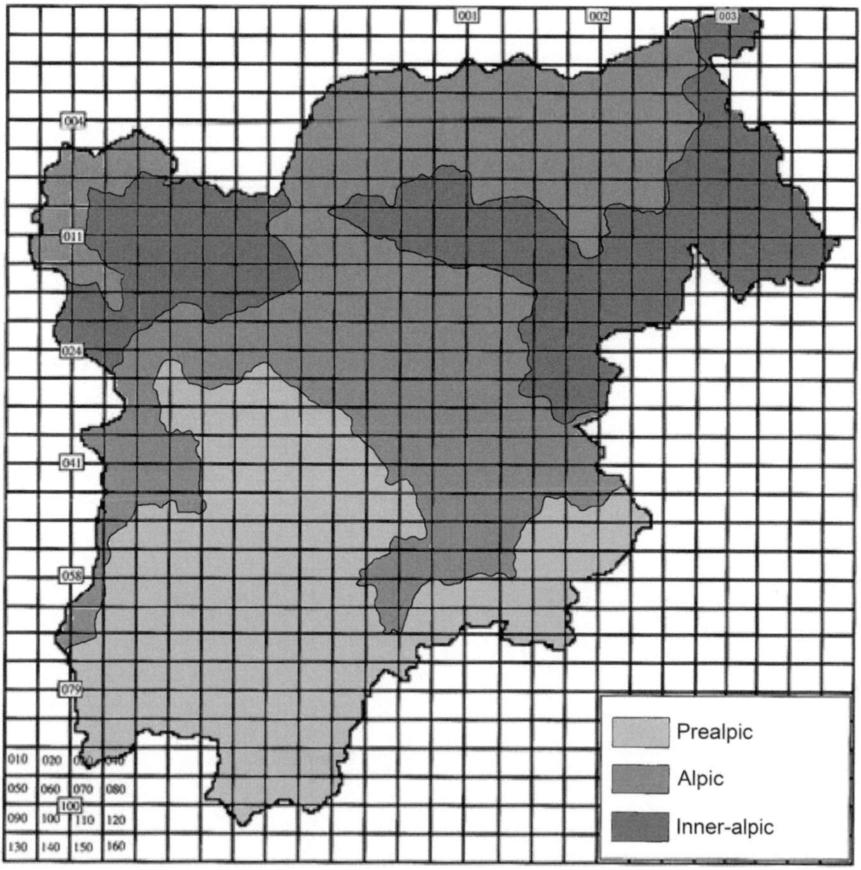

Fig. 3.6 Mao of the hygric-pluvial continentality types; the map is referred to the UTM network (from Gafta and Pedrotti 1998, modified)

represented on a map on a scale of 1:250,000 (Pedrotti and Gafta 2003). Each megageoseries differs from the others at least by a vegetation series; the gradients of the megageoseries, therefore, are "partial" gradients.

3.5 List of the Series Identified

The vegetation series (or sigmeta) are listed according to altitudinal belts; within each altitudinal belt, the series have been grouped according to the ecological and floristic-physiognomic affinity of the association that constitutes the series head.

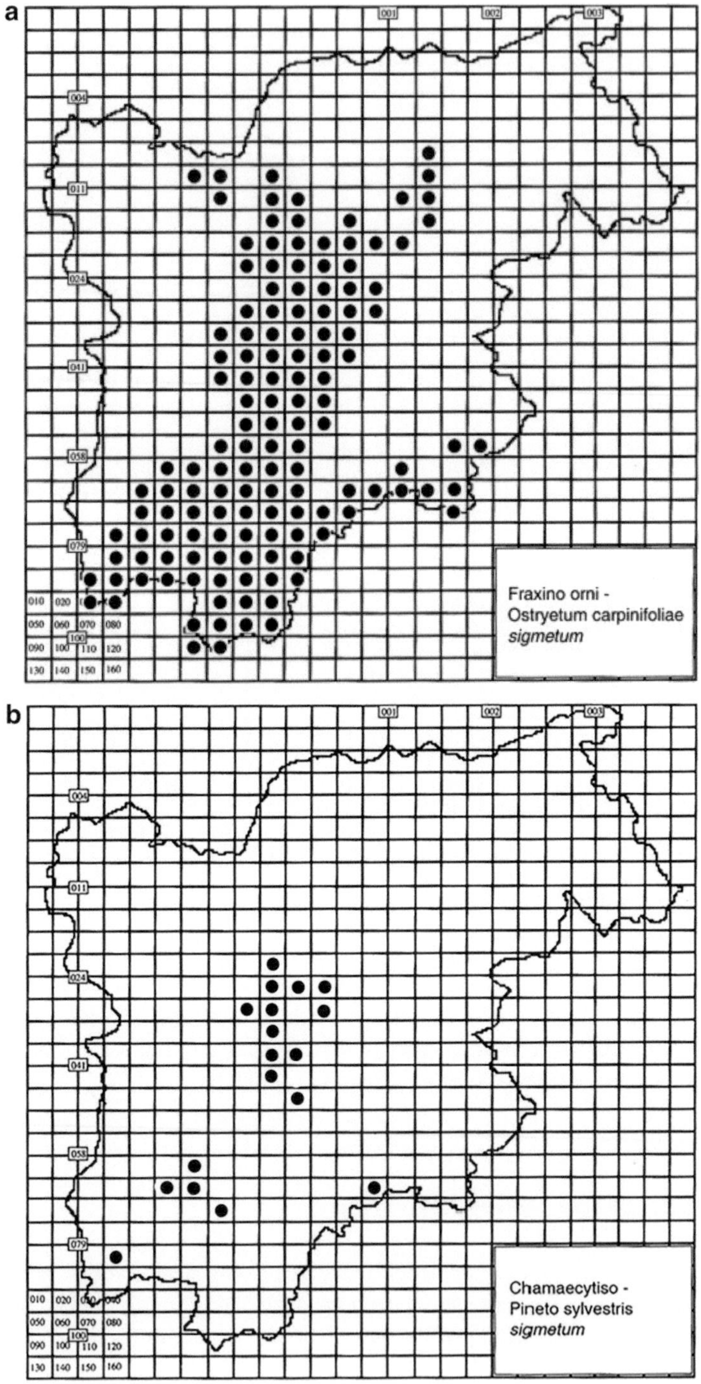

Fig. 3.7 (**a**) Distribution of *Fraxino orni-Ostryetum carpinifoliae* sigmetum in the Trentino-Alto Region referred to the UTM network. (**b**) Distribution of *Chamaecytiso-Pineto sylvestris* sigmetum in the Trentino-Alto Region referred to the UTM network

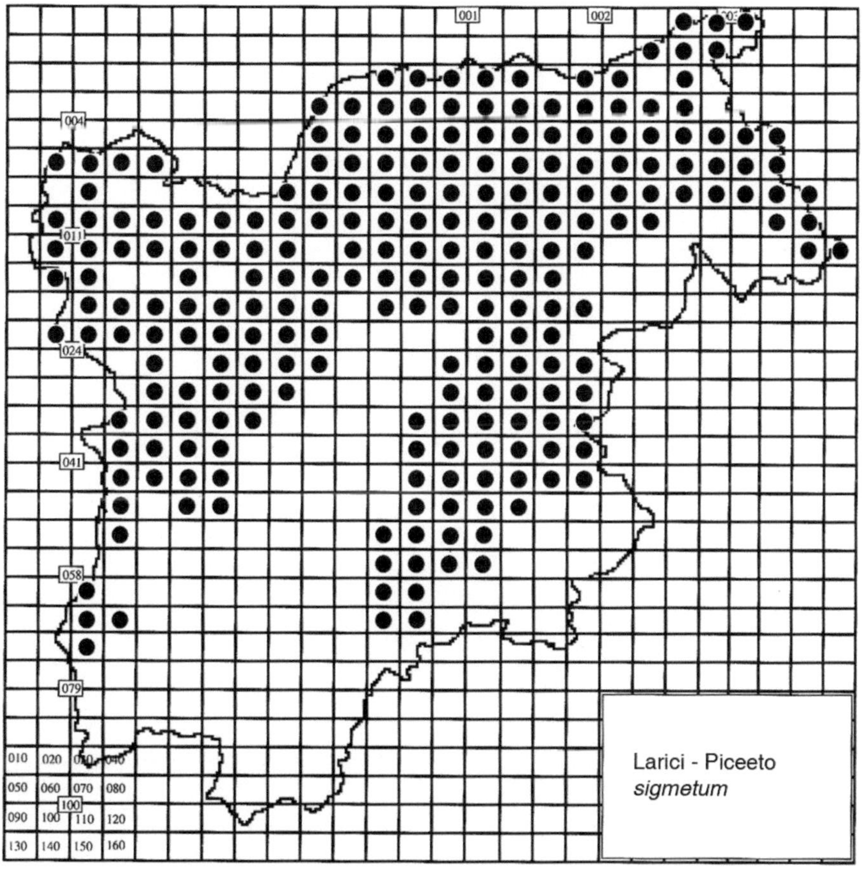

Fig. 3.8 Distribution of *Larici-Piceeto* sigmetum in the Trentino-Alto Region referred to the UTM network

3.5.1 Zonal, Intrazonal and Extrazonal Series

Nival Belt

(1) Cryo-orotemperate, alpic, humid, acidophilous sigmion of *Andraea nivalis* [*Andraeion nivalis* sigmion]. Alpic, inner-alpic, metamorphic-magmatic.
(2) Cryo-orotemperate, alpic, humid, basophilus sigmion of *Andreaea rupestris* [*Andraeion rupestris* sigmion]. Alpic, inner-alpic, calcareous-dolomitic.
(3) Cryo-orotemperate, alpic, humid, acidophilous sigmion of *Androsace alpina* [*Androsacion alpinae* sigmion]. Alpic, inner-alpic, metamorphic-magmatic.
(4) Cryo-orotemperate, alpic, humid, neutrophilous sigmion of *Draba hoppeana* [*Drabion hoppeanae* sigmion]. Alpic, inner-alpic, calcareous-schistose.
(5) Cryo-orotemperate, alpic, humid, basophilus sigmion of *Thlaspi rotundifolium* [*Thlaspion rotundifolii* sigmion]. Prealpic, alpic, inner-alpic, calcareous-dolomitic.

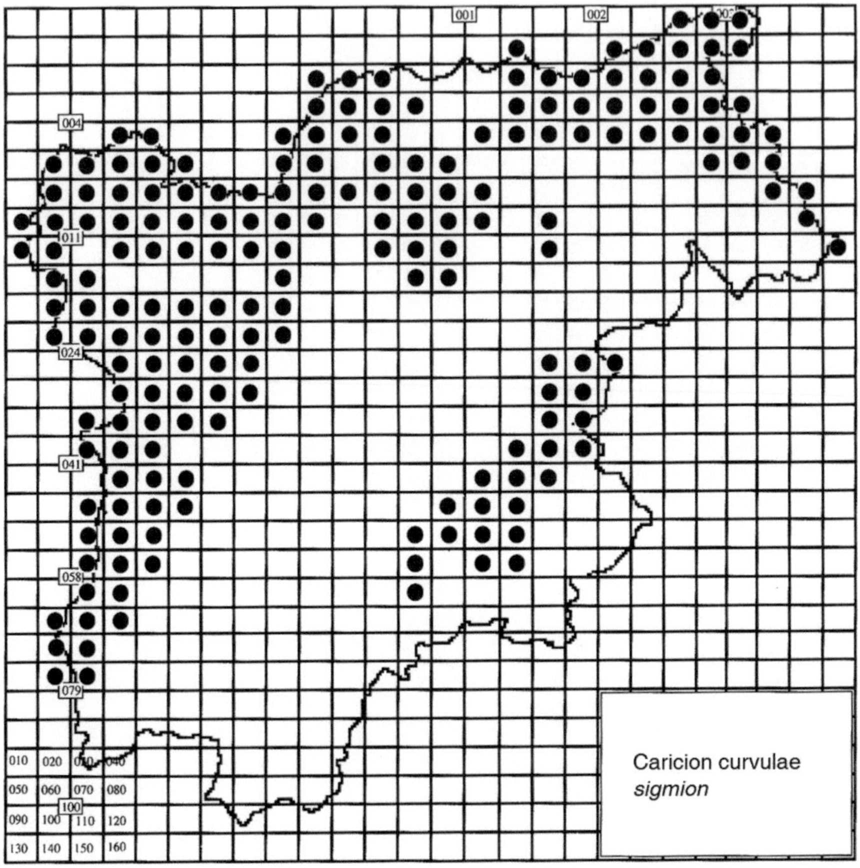

Fig. 3.9 Distribution of *Caricion curvulae* sigmion in the Trentino-Alto Region referred to the UTM network

Alpine Belt

(6) Cryo-orotemperate, alpic, humid basophilus sigmion of *Arabis caerulea* [*Arabidion coeruleae* sigmion]. Intrazonal (*Caricion firmae* sigmion), prealpic, alpic, inner-alpic, calcareous-dolomitic.
(7) Cryo-orotemperate, alpic, humid, series of *Salix herbacea* [*Saliceto herbaceae* sigmetum]. Intrazonal (*Caricion curvulae* sigmion), alpic, inner-alpic, metamorphic-magmatic.
(8) Cryo-orotemperate, alpic, humid, acidophilous series of C*arex curvula* [C*ariceto curvulae* sigmetum]. Alpic, inner-alpic, metamorphic-magmatic (Fig. 3.9).

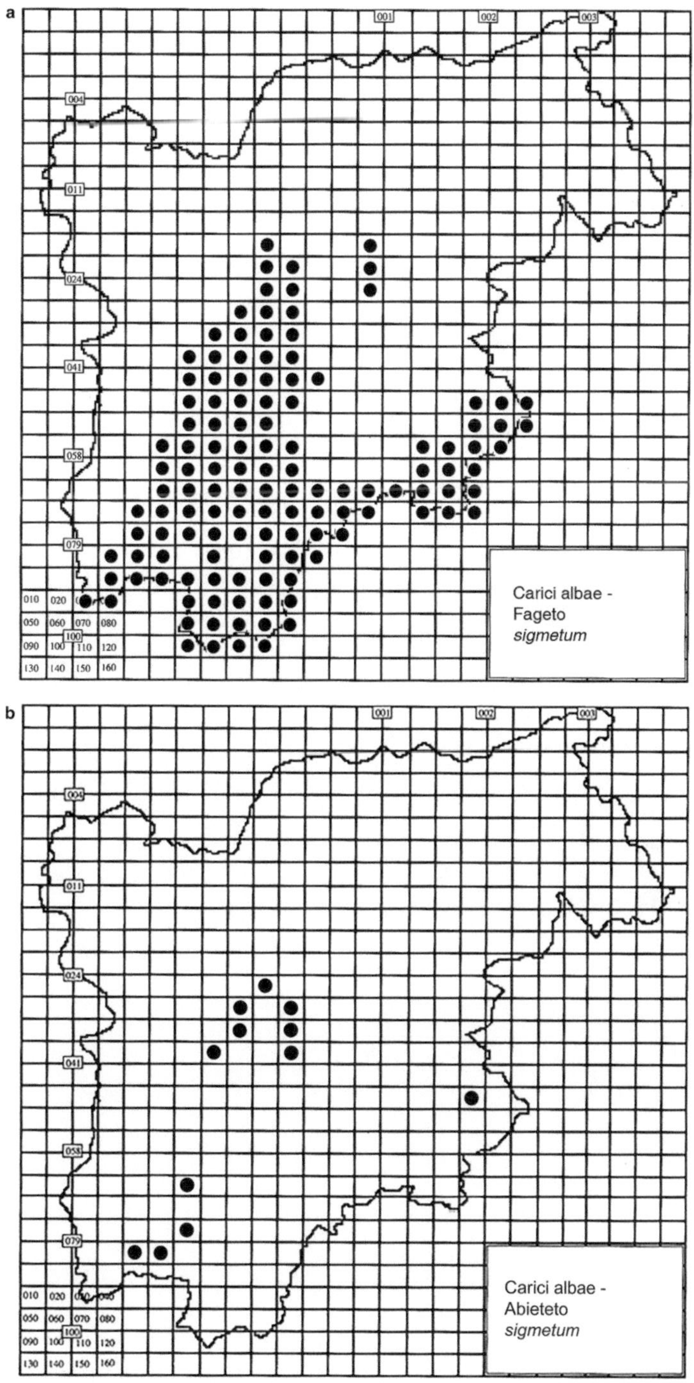

Fig. 3.10 (**a**) Distribution of *Carici albae-Fageto* sigmetum in the Trentino-Alto Region referred to the UTM network. (**b**) Distribution of *Carici albae-Abieteto* sigmetum in the Trentino-Alto Region referred to the UTM network

(9) Cryo-orotemperate, centre-south-alpic, humid, acidophilous series of *Carex curvula* and *Primula daonensis* [*Primulo-Cariceto curvulae* sigmetum]. Alpic, metamorphic-magmatic, Ortles-Cevedale, Adamello.
(10) Cryo-orotemperate, centre-south-alpic, humid, acidophilous series of *Saponaria pumilio* [*Senecio-Saponarieto pumilae* sigmetum]. Alpic, metamorphic-magmatic, Lagorai.
(11) Cryo-orotemperate, alpic, humid, acidophilous series of *Festuca halleri* [*Festuceto halleri* sigmetum]. Alpic, inner-alpic, metamorphic-magmatic.
(12) Cryo-orotemperate, east-alpic, humid acidophilous series of *Festuca varia* [*Pulsatillo albae-Festuceto variae* sigmetum]. Alpic, metamorphic-magmatic.
(13) Cryo-orotemperate, alpic, humid, acidophilous series of *Festuca scabriculmis* [*Festuceto scabriculmis* sigmetum]. Alpic, metamorphic-magmatic.
(14) Cryo-orotemperate, east-alpic, humid, neutrophilous series of *Elyna myosuroides* [*Elyneto myosuroidis* sigmetum]. Alpic, inner-alpic, calcareous-dolomitic.
(15) Cryo-orotemperate, alpic, humid, neutro-acidophilus series of *Carex rosae* [*Elyno-Cariceto rosae* sigmetum]. Intrazonal (*Elyno-Oxytropido* sigmion), prealpic, alpic, calcareous-dolomitic (Erschbamer 1992).
(16) Cryo-orotemperate, alpic, humid, basophilus series of *Carex rupestris* [*Cariceto rupestris* sigmetum]. Alpic, inner-alpic, calcareous-dolomitic.
(17) Cryo-orotemperate, alpic, humid, basophilus series of *Carex mucronata* [*Cariceto mucronatae* sigmetum]. Alpic, inner-alpic, calcareous-dolomitic.
(18) Cryo-orotemperate, alpic, humid, basophilus series of *Carex firma* [*Cariceto firmae* sigmetum]. Prealpic, alpic, inner-alpic, calcareous-dolomitic.
(19) Cryo-orotemperate, alpic, humid, basophilus series of *Dryas octopetala* [*Dryadeto octopetalae* sigmetum]. Prealpic, alpic, inner-alpic, calcareous-dolomitic.
(20) Cryo-orotemperate, alpic, humid, basophilus series of *Sesleria varia* [*Seslerio-Cariceto sempervirentis* sigmetum]. Prealpic, alpic, inner-alpic, calcareous-dolomitic.
(21) Cryo-orotemperate, alpic, humid, basophilus series of *Trifolium thalii* [*Trifolio thalii-Festucetum nigricantis* sigmetum]. Prealpic, alpic, inner-alpic, calcareous-dolomitic.
(22) Cryo-orotemperate, alpic, humid, basophilus series of *Carex ferruginea* [*Cariceto ferrugineae* sigmetum]. Prealpic, alpic, inner-alpic, calcareous-dolomitic.

Subalpine Belt

(23) Cryo-orotemperate, alpic, humid, acidophilous series of *Vaccinium gaultherioides* [*Empetro-Vaccinieto gaultherioidis* sigmetum]. Alpic, inner-alpic, metamorphic-magmatic.
(24) Cryo-orotemperate, alpic, humid, neutrophilous series of *Loiseleuria procumbens* [*Loiseleurio-Cetrarieto* sigmetum]. Intrazonal (*Empetro-Vaccinieto gaultherioidis* sigmetum), alpic, inner-alpic, metamorphic-magmatic.

(25) Orotemperate, alpic, humid acidophilous series of *Rhododendron ferrugineum* [*Rhododendreto ferruginei* sigmetum]. Alpic, inner-alpic, metamorphic-magmatic.
(26) Orotemperate, alpic, humid, acidophilous mesic series of *Larix decidua* [*Rhododendro-Lariceto deciduae* sigmetum]. Alpic, metamorphic magmatic.
(27) Orotemperate, centre-east-alpic, humid, basophilus series of *Erica carnea* [*Ericeto carneae* sigmetum]. Prealpic, alpic, calcareous-dolomitic.
(28) Orotemperate, centre-east-alpic, humid, basophilus of *Rhododendron hirsutum* [*Rhododendreto hirsuti* sigmetum]. Prealpic, alpic, inner-alpic, calcareous-dolomitic.
(29) Orotemperate, centre-east-alpic, humid, acidophilous series of *Pinus mugo* and *Rhododendron ferrugineum* [*Rhododendro ferruginei-Pineto prostratae* sigmetum]. Prealpic, alpic, inner-alpic, metamorphic-magmatic.
(30) Orotemperate, centre-east-alpic, humid, acidophilous series of *Pinus mugo* [*Vaccinio-Pineto mugo* sigmetum]. Alpic, metamorphic-magmatic.
(31) Orotemperate, alpic, humid, basophilus, hygrophilous series of *Pinus mugo* [*Rhododendro hirsuti-Pineto prostratae* sigmetum]. Prealpic, alpic, inner-alpic, calcareous-dolomitic.
(32) Orotemperate, alpic, humid, basophilus, xeric series of *Pinus mugo* [*Erico carneae-Pineto prostratae* sigmetum]. Alpic, calcareous-dolomitic.
(33) Orotemperate, centre-south-alpic, humid, acidophilous series of *Pinus mugo* and *Sorbus chamaemespilus* [*Sorbo chamaemespili-Pineto mugo* sigmetum]. Intrazonal (*Erico-Pinion mugo* sigmion), prealpic, alpic, calcareous-dolomitic, decalcified and acidified soil.
(34) Orotemperate, alpic, humid, acidophilous series of *Pinus cembra* [*Larici-Pineto cembrae* sigmetum]. Alpic, metamorphic-magmatic.
(35) Orotemperate, alpic, humid, basophilus series, of *Larix decidua* [*Lariceto deciduae* sigmetum]. Alpic, calcareous-dolomitic.
(36) Orotemperate, alpic, humid, basophilus series of *Pinus cembra* [*Pineto cembrae* sigmetum]. Alpic, calcareous-dolomitic.
(37) Orotemperate, alpic, humid, acidophilous series of *Juniperus nana* [*Junipero nanae-Arctostaphyleto* sigmetum]. Alpic, metamorphic-magmatic.
(38) Orotemperate, alpic, humid, acidophilous series of *Pinus cembra* [*Cotoneastro integerrimae-Pineto cembrae* sigmetum]. Alpic, metamorphic-magmatic.
(39) Orotemperate, alpic, humid, acidophilous, edafo-xerophilous series of *Larix decidua* [*Junipero nanae-Lariceto deciduae* sigmetum]. Alpic, metamorphic-magmatic.
(40) Orotemperate, alpic, humid, series of *Alnus viridis* [*Alneto viridis* sigmetum]. Intrazonal (*Rhododendro ferruginei* sigmetum, *Larici-Pineto cembrae* sigmetum, *Larici-Piceeto* sigmetum). Alpic, inner-alpic (Fig. 3.8).
(41) Orotemperate, alpic, humid, acidophilous series of *Salix helvetica* [*Saliceto helveticae* sigmetum]. Intrazonal (*Rhododendreto ferruginei* sigmetum). Alpic, metamorphic-magmatic, Paradiso del Cevedale (Val Martello).
(42) Orotemperate, centre-east-alpic, humid, basophilus series of *Salix caesia* and *Salix arbuscula* [*Saliceto caesio-arbusculae* sigmetum]. Intrazonal (*Erico-Pinion*

mugo sigmion, *Adenostylo-Piceeto* sigmetum), prealpic, calcareous-dolomitic, Gruppo di Brenta.
(43) Orotemperate, alpic, humid, series of *Salix waldsteiniana* [*Saliceto waldsteinianae* sigmetum]. Intrazonal (*Erico-Pinion mugo* sigmion), alpic, calcareous-dolomitic, Val Venegia (Boiti et al. 1989).

Montane Belt

(44) Lower supratemperate, European, humid, acidophilous series, of *Pinus sylvestris* [*Vaccinio vitis-idaeae-Pineto sylvestris* sigmetum]. Extrazonal, prealpic, alpic, inner-alpic, metamorphic-magmatic.
(45) Supratemperate, centre-south-alpic, acidophilous, xeric and subxeric series, of *Pinus sylvestris* [*Astragalo vesicarii-Pineto sylvestris* sigmetum]. Inner-alpic, metamorphic and magmatic-intrusive, Val Venosta (Braun-Blanquet 1961).
(46) Lower supratemperate, centre-south-alpic, subhumid and humid, acidophilous series of *Pinus sylvestris* and *Anthericum liliago* [*Antherico liliaginis-Pineto sylvestris* sigmetum]. Intrazonal (*Luzulo niveae-Querceto petraeae* sigmetum), prealpic, alpic, inner-alpic, metamorfic-magmatic, Val Pusteria and Val di Cembra.
(47) Lower supratemperate, centre-east-alpic, humid, basophilus series of *Pinus sylvestris* [*Erico-Pineto sylvestris* sigmetum]. Intrazonal (*Carici albae-Fageto sylvaticae* sigmetum, *Adenostylo glabrae-Piceeto* sigmetum), alpic, inner-alpic, calcareous-dolomitic.
(48) Lower supratemperate, alpic, humid, neutrophilous series of *Pinus sylvestris* [*Molinio litoralis-Pineto sylvestris* sigmetum]. Intrazonal (*Carici abae-Fageto* sigmetum, *Adenostylo glabrae-Piceeto* sigmetum), prealpic, alpic, calcareous-marl, Val d'Algone and Val Badia (Minghetti 2003).
(49) Upper supratemperate, alpic, humid, acidophilous xeric series of *Picea abies* [*Larici-Piceeto* sigmetum]. Alpic, metamorphic-magmatic.
(50) Upper supratemperate, alpic, humid, edafo-hygrophylous acidophilous series of *Picea abies* and *Listera cordata* [*Listero-Piceeto* sigmetum]. Intrazonal (*Larici-Piceeto* sigmetum), alpic, inner-alpic, metamorphic-magmatic.
(51) Lower supratemperate, alpic, humid, acidophilous series of *Picea abies* [*Luzulo nemorosae-Piceeto abietis* sigmetum]. Alpic, inner-alpic, magmatic effusive.
(52) Supratemperate, alpic, humid, basophilus series of *Picea abies* [*Adenostylo glabrae-Piceeto albae* sigmetum]. Prealpic, alpic, inner-alpic, calcareous-dolomitic.
(53) Supratemperate and lower supratemperate, alpic, humid, acidophilous, rocky series of *Picea excelsa* [*Asplenio-Piceeto* sigmetum]. Intrazonal (*Larici-Piceeto* sigmetum, *Luzulo nemorosae-Piceeto* sigmetum), alpic, inner-alpic, metamorphic-magmatic.
(54) Lower supratemperate, east-alpic, humid, series of *Fagus sylvatica* [*Galio odorati-Fageto sylvaticae* sigmetum]. Metamorphic-magmatic alpic, prealpic.

(55) Upper supratemperate, east-alpic, humid, series of *Fagus sylvatica* [*Calamagrostio villosae-Fageto sylvaticae* sigmetum]. Metamorphic-magmatic, alpic, prealpic.
(56) Lower supratemperate, east-alpic, humid, acidophilous series of *Fagus sylvatica* [*Luzulo niveae-Fageto sylvaticae* sigmetum]. Metamorphic-magmatic, prealpic, alpic.
(57) Lower supratemperate, east-alpic, humid, basophilus series of *Fagus sylvatica* [*Carici albae-Fageto sylvaticae* sigmetum]. Prealpic, alpic, calcareous-dolomitic (Fig. 3.10a).
(58) Lower supratemperate, prealpic, humid, subacidophilous series of *Fagus sylvatica* [*Cardamino pentaphylli-Fageto sylvaticae* sigmetum]. Prealpic.
(59) Lower supratemperate, east-alpic, humid, basophilus rocky series of *Fagus sylvatica* and *Sesleria varia* [*Seslerio albicantis-Fageto* sigmetum]. Intrazonal (*Cardamino pentaphylli-Fageto* sigmetum), prealpic, calcareous-dolomitic, Gruppo di Brenta, southern slopes.
(60) Lower supratemperate, alpic, humid, acidophilous series of *Abies alba* [*Vaccinio-Abieteto albae* sigmetum]. Intrazonal (*Luzulo-Piceeto* sigmetum, *Larici-Piceeto* sigmetum), alpic, silicatic.
(61) Lower supratemperate, alpic, humid, basophilus series of *Abies alba* and *Carex alba* [*Carici albae-Abieteto albae* sigmetum]. Intrazonal (*Carici albae-Fageto* sigmetum), prealpic, calcareous-dolomitic (Fig. 3.10b).
(62) Lower supratemperate, alpic, humid, acidophilous series of *Abies alba* and *Orthilia secunda* [*Pyrolo-Abieteto* sigmetum]. Intrazonal (*Cardamino pentaphylli-Fageto* sigmetum), prealpic calcareous, decalcified and acidified soil, altipiani di Serrada, Folgaria, Lavarone, sopra Grigno (Gafta 1994).
(63) Lower supratemperate, alpic, humid, acidophilous series of *Abies alba* and *Galium odoratum* [*Galio odorati-Abieteto albae* sigmetum]. Intrazonal (*Galio odorati-Fageto* sigmetum), prealpic, alpic, silicatic.
(64) Lower supratemperate, alpic, humid, acidophilous series of *Abies alba* and *Luzula nivea* [*Luzulo niveae—Abieteto albae* sigmetum]. Intrazonal (*Luzulo niveae-Fageto* sigmetum), prealpic, alpic, silicatic.
(65) Lower supratemperate, alpic, humid, basophilus series of *Abies alba* [*Cardamino pentaphylli-Abieteto* albae sigmetum]. Intrazonal (*Cardamino pentaphylli-Fageto* sigmetum), prealpic, calcareous-dolomitic.
(66) Supratemperate, alpic, humid, basophilus series of *Abies alba* [*Adenostylo glabrae-Abieteto albae* sigmetum]. Intrazonal (*Adenostylo glabrae-Piceetum* and *Cardamino pentaphylli-Fageto* sigmetum), prealpic, calcareous-dolomitic.
(67) Upper supratemperate, alpic, humid, acidophilous series of *Abies alba* [*Calamagrostio villosae-Abieteto albae* sigmetum]. Intrazonal (*Calamagrostio villosae-Fageto* sigmetum), prealpic, alpic, silicatic.
(68) Lower supratemperate, alpic, humid, basophilus series of *Taxus baccata* [*Taxo-Fageto sylvaticae* sigmetum]. Intrazonal (*Carici albae-Fageto* sigmetum), prealpic, calcareous-dolomitic.
(69) Lower supratemperate, centre-south-alpic, humid, edapho-hygrophilous series of *Acer pseudoplatanus* [*Calamintho grandiflorae-Acereto pseudoplatani*

sigmetum]. Intrazonal (*Galio odorati-Fageto* sigmetum and *Cardamino pentaphylli-Fageto* sigmetum), prealpic, Adamello.
(70) Upper and lower supratemperate, alpic, humid, acidophilous series of *Acer pseudoplatanus* and *Fraxinus excelsior* [*Aceri pseudoplatani-Fraxineto excelsioris* sigmetum]. Intrazonal (*Fraxino orni-Ostryeto carpinifoliae* sigmetum, *Vaccinio-Pineto sylvestris* sigmetum, *Luzulo-Piceeto* sigmetum), alpic and inner-alpic, gola di Racines, gola Tina,Valsura, Val Sarentina, Val d'Ega, Val Gardena (Peer 1995).
(71) Upper and lower supratemperate, alpic, humid, acidophilous series of *Acer pseudoplatanus* and *Fagus sylvatica* [*Aceri-Fageto* sigmetum]. Intrazonal (*Fraxino orni-Ostryeto carpinifoliae* sigmetum, *Vaccinio-Pineto sylvestris* sigmetum, *Luzulo-Piceeto* sigmetum), alpic, inner-alpic. Peer (1995) also reported the *Aceri-Tilietum* association, *Aceri-Tilieto* sigmetum.

Colline Belt

(72) Mesotemperate southalpine, subhumid, acidophilous series of *Quercus pubescens* [*Arabidi turritae-Querceto pubescentis* sigmetum]. Alpic and inner-alpic, metamorphic-magmatic.
(73) Mesotemperate and upper mesotemperate, southalpine, humid and subhumid, prealpic, acidophilous series of *Quercus petraea* [*Luzulo niveae-Quercetum petraeae* sigmetum]. Alpic, inneralpic, metamorphic-magmatic, Val di Cembra, Val Pusteria, Val Venosta, Val di Genova, Val di Sole (Pedrotti 2015).
(74) Mesotemperate and upper mesotemperate, southalpine, acidophilous series of *Carpinus betulus* [*Galio laevigati-Carpinetum betuli* sigmetum]. Prealpic, metamorphic-magmatic, Gocciadoro near Trento, Valsugana, Val Rendena, Valle del Chiese (Fig. 3.11).
(75) Upper mesotemperate, southalpine, humid, acidophilous series of *Fraxinus excelsior* [*Salvio glutinosae-Fraxineto excelsioris* sigmetum]. Alpic and inneralpic, metamorphic-magmatic.
(76) Upper mesotemperate, southalpine, subhumid, basophilus series of *Fraxinus excelsior* [*Fraxineto excelsioris* s.l. sigmetum]. Prealpic, calcareous.
(77) Lower supratemperate, southalpine, humid, acidophilous series of *Quercus cerris* [*Querceto cerris* s.l. sigmetum]. Prealpic, metamorphic-magmatic, Valle del Chiese.
(78) Upper mesotemperate, southalpine, subhumid, basophilus series of *Ostrya carpinifolia* [*Fraxino orni-Ostryeto carpinifoliae* sigmetum]. Prealpic, calcareous (Fig. 3.7a).
(79) Upper mesotemperate, southalpine, subhumid, basophilus series of *Ostrya carpinifolia* and *Buglossoides purpurocaerulea* [*Buglossoido-Ostryeto carpinifoliae* sigmetum]. Prealpic, calcareous.
(80) Upper mesotemperate, southalpine, subhumid, basophilus series of *Ostrya carpinifolia* and *Sesleria varia* [*Seslerio albicantis-Ostryeto carpinifoliae* sigmetum]. Prealpic, calcareous.

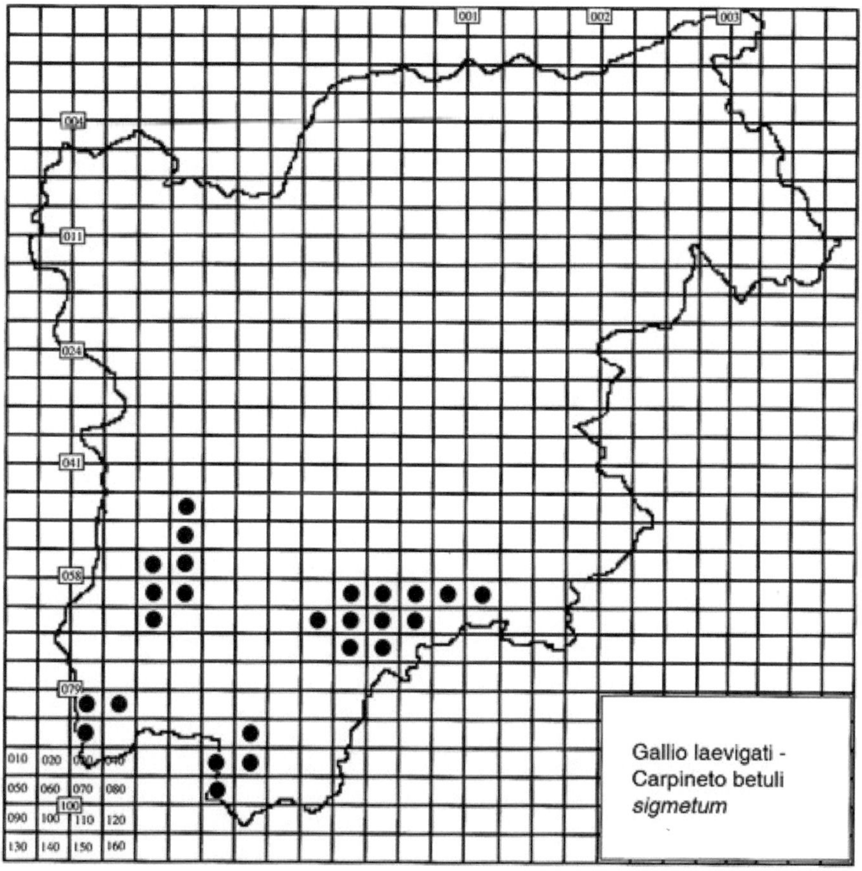

Fig. 3.11 Distribution of *Galio laevigati-Carpineto betu*li sigmetum in the Trentino-Alto Region referred to the UTM network

(81) Mesotemperate, southalpine, humid, basophilus series of *Pinus sylvestris* [*Chamaecytiso pupurei-Pineto sylvestris* sigmetum]. Intrazonal (*Fraxino orni-Ostryeto carpinifoliae* sigmetum), prealpic, calcareous (Fig. 3.7b).
(82) Upper mesotemperate, southalpine, humid, basophilus series of *Pinus mugo* and *Amelanchier ovalis* [*Amelanchiero-Pineto mugo* sigmetum]. Intrazonal (*Fraxino orni-Ostryeto carpinifoliae* sigmetum), prealpic, calcareous, Valsugana (Minghetti 1996).
(83) Lower mesotemperate, southalpine, subhumid, basophilus series of *Quercus ilex* [*Celtidi australis-Querceto ilicis* sigmetum]. Extrazonal, prealpic, calcareous, Lago di Garda, Valle del Sarca, Valle dell'Adige between Domegliara and Valle Aviana (Fig. 3.12).

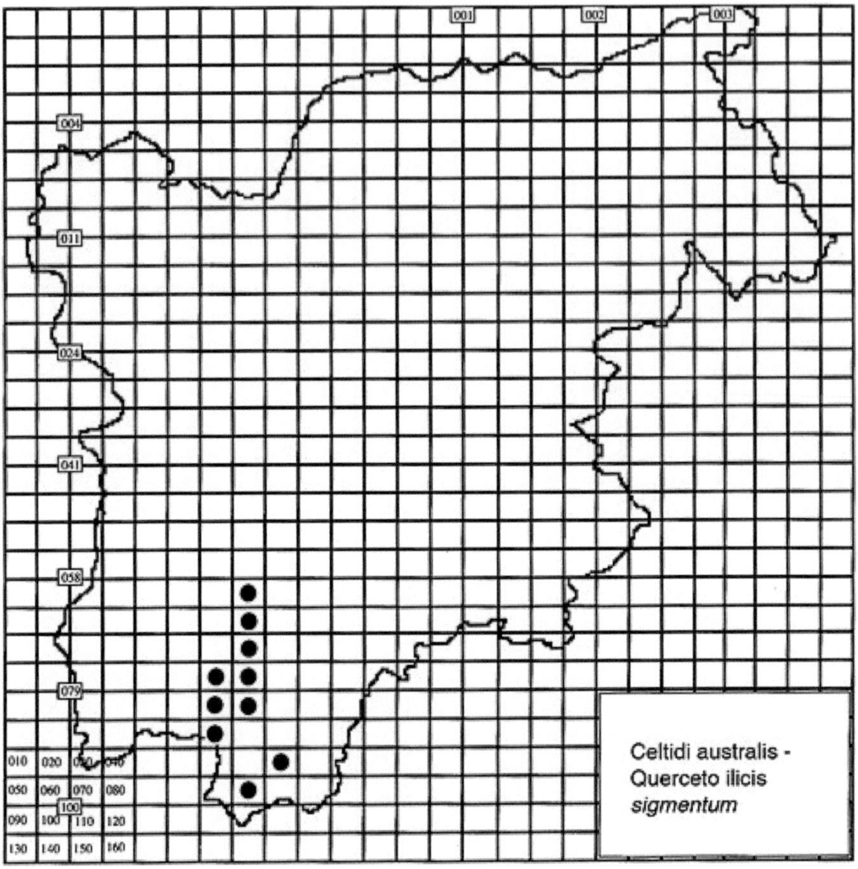

Fig. 3.12 Distribution of *Celtidi australis-Querceto ilicis* sigmetum in the Trentino-Alto Region referred to the UTM network

Azonal Series

(84) Mesotemperate and supratemperate, European, humid and subhumid, peaty-swampy series of S*alix cinerea* [S*aliceto cinereae* sigmetum]. Prealpic, alpic, inner-alpic.
(85) Montane, European, humid and subhumid, peaty-swampy series of *Salix pentandra* [*Saliceto pentandrae* sigmetum s.l.]. Prealpic, alpic, torbiera di Fiavé, Lago della Serraia.
(86) Upper supratemperate, humid and subhumid, peaty-swampy series of *Alnus incana* and *Salix pentandra* [*Alno-Saliceto pentandrae* sigmetum]. Inner-alpic, S. Valentino alla Muta (Val Venosta) (Pedrotti and Gafta 1996).

(87) Lower supratemperate, European, humid and subhumid, peaty-swampy series of *Alnus glutinosa* and *Carex elongata* [*Carici elongatae-Alneto glutinosae* sigmetum]. Prealpic, Lago della Serraia.
(88) Lower supratemperate, European, humid and subhumid, peaty-swampy series of *Alnus glutinosa* and *Carex elata* [*Carici elatae-Alneto glutinosae* sigmetum]. Prealpic, Laghestel di Piné, Monte Barco (Pedrotti 2004).
(89) Upper mesotemperate xeric and lower supratemperate suxeric, European, humid and subhumid, peaty-swampy series of *Alnus glutinosa* and *Caltha palustris* [*Caltho-Alnetum glutinosae* sigmetum]. Inner-alpic, alpic, Val Venosta.
(90) Upper mesotemperate, European, humid and subhumid, peaty-swampy series of *Alnus glutinosa* and *Carex acutiformis* [*Carici acutiformis-Alneto glutinosae* sigmetum]. Prealpic, Roncegno and Resenzuola (Valsugana), Lago d'Idro (Valle del Chiese).
(91) Upper mesotemperate, European, humid and subhumid, peaty-swampy series of *Alnus glutinosa* and *Thelypteris palustris* [*Thelypteridi-Alneto glutinosae* sigmetum]. Prealpic, Laghi di Madrano, Caldonazzo and Levico.
(92) Supratemperate, European, humid, peaty series of *Pinus sylvestris* and *Molinia coerulea* [*Molinio coeruleae-Pineto sylvestris* sigmetum]. Prealpic, alpic.
(93) Lower supratemperate, European, humid, peaty series of *Pinus sylvestris* and *Vaccinium uiginosum* [*Vaccinio uliginosi-Pinetum sylvestris* sigmetum]. Alpic, Altopiani del Renon, Nova Ponente e Nova Levante (Minghetti 2003).
(94) Lower supratemperate, European, humid, peaty series, of *Betula pubescens* [*Vaccinio uliginosi-Betuleto pubescentis* sigmetum]. Altopiano di Barco, prealpic.
(95) Orotemperate, European, humid, peaty series of *Picea abies* and *Sphagnum girgensohnii* [*Sphagno girgensohnii-Piceeto* sigmetum]. Torbiera Pezzabosco, Passo di Carlomagno, varie località in Alto Adige, prealpic, alpic.
(96) Orotemperate, European, humid, peaty series of *Pinus mugo* [*Sphagno-Pineto mugo* sigmetum]. Prealpic, alpic, inner-alpic.
(97) Supratemperate and orotemperate, alpic, humid and subhumid, alluvial series of *Myricaria gemanica* [*Salici-Myricarieto* sigmetum]. Inner-alpic, alpic, Prato allo Stelvio (Pedrotti et al. 1974), Val di Fiemme (Pedrotti and Gafta 1996), Val Sarentino (Kiem 1992).
(98) Supratemperate and orotemperate, alpic, humid, alluvial series of *Hippophaë rhamnoides* [*Salici incanae-Hippophaëto* sigmetum]. Inner-alpic, alpic.
(99) Mesotemperate and supratemperate, alpic, humid and subhumid, alluvial series of *Salix incana* [*Saliceto incano-purpureae* sigmetum]. Inner-alpic, alpic, prealpic.
(100) Mesotemperate and supratemperate, European, humid and subhumid, alluvial series of *Salix alba* [*Saliceto albae* sigmetum]. Inner-alpic, alpic, prealpic.
(101) Supratemperate, alpic, humid, alluvial series of *Alnus incana* [*Alneto incanae* sigmetum]. Prealpic, alpic, inner-alpic.
(102) Supratemperate, alpic, suboceanic and subcontinental, alluvial series of *Alnus glutinosa* [*Stellario nemorum-Alneto glutinosae* sigmetum]. Alpic.
(103) Mesotemperate, European, alluvial series of *Fraxinus excelsior* and *Carex remota* [*Carici remotae-Fraxineto excelsioris* sigmetum]. Prealpic.

(104) Supratemperate, European, alluvial series of *Fraxinus excelsior* and *Prunus padus* [*Pruno padi-Fraxineto excelsioris* sigmetum]. Alpic, Val Pusteria (Pedrotti and Gafta 1996).
(105) Mesotemperate, humid, alluvial series of *Populus alba* [*Populeto albae* sigmetum l.s.]. Val d'Adige.
(106) Mesotemperate, humid, alluvial series of *Quercus robur* [*Querceto roboris* s.l.]. Reported by Schiechtl et Stern (1976) in Val d'Isarco, it probably no longer exists today, because of reclamation works in the valley bottom.
(107) Lower temperate, alpic, humid, alluvial, basophilus series of *Pinus sylvestris* and *Salix elaeagnus* [*Salici elaeagni-Pineto sylvestris* sigmetum]. Prealpic, alpic, calcareous-dolomitic, Val di Landro, Val Badia, Lago di Tovel, Val d'Algone.
(108) Mesotemperate, alpic, humid, alluvial, basophilus series of *Pinus sylvestris* and *Alnus incana* [*Alno-Pineto sylvestris* sigmetum]. Prealpic, calcareous-dolomitic, Valsugana, Val d'Algone.

Aquatic Vegetation (Lakes, Streams, Springs, and Other Types)

(109–116) Series of lakes (*Phragmition communis* sigmion, *Magnocaricion* sigmion, *Cirsio brachycephali-Bolboschoenion* sigmion, *Nymphaeion albae* sigmion, *Potamion pectinati* sigmion, *Ranunculion fluitantis* sigmion, *Nanocyperion* sigmion, *Eleocharition acicularis* sigmion).
(117–120) Series of springs and streams (*Cardamino-Montion* sigmion, *Cratoneurion* sigmion, *Adiantion* sigmion, *Glycerio-Sparganion* sigmion).
(121–122) Series of other aquatic environments (*Phalaridion arundinaceae* sigmion, *Oenanthion aquaticae* sigmion).

Torficolous Vegetation (Mires)

(123–128) Series of mires (*Caricion fuscae* sigmion, *Caricion davallianae* sigmion, *Rhynchosporion albae* sigmion, *Caricion lasiocarpae* sigmion, *Caricion atrofusco-saxatilis* sigmion, *Sphagnion medii* sigmion).

3.5.2 Other Possible Series

In Alto Adige two vegetation series should be sought out, characterized by *Pinus uncinata* and *Pinus engadinensis* respectively, with the *Erico-Pinetum uncinatae* and *Carici humilis-Pinetum engadinensis* associations. They were observed in Livigno, in Italy, in an area on the Swiss border, the Valle del Spöll; these two associations were reported for this location by Braun-Blanquet (1946) and Reinalter (2003).

3.6 General Considerations on the Vegetation

The vegetation landscape of the Trentino-Alto Adige Region is characterized by 128 vegetation series. This evaluation is incomplete, because in some cases the sigmion (which can also comprise more than one sigmetum) was indicated. In addition, the presence and nomenclature of some series should be clarified. Finally, the vegetation series of humd environments (lakes, swamps and mires) should be added, because for the moment they were only indicated as sigmion, and were not calculated. For a description of the series, see Pedrotti (2006, 2010).

This high number is the consequence of multiple causes, such as latitudinal extension, lithological substrates, geomorphological conditions and phytoclimate. It demonstrates the complexity and the great phytogeographic diversity of the territory under study, with a very marked gradient, passing from the prealpic to the inner-alpic sector, which can be highlighted by referring to both the flora (including the endemisms) and the vegetation series and megageoseries, as will be illustrated below.

Of the 128 vegetation series catalogued to date, 42 are characterized by herbaceous and cryptogam species, 20 by frutescent species and 66 by arboreal species. Of the 23 species characterized by shrubs, 15 are of the subalpine belt and 7 are azonal: *Pinus mugo* 7 series (5 zonal, 1 intrazonal, 1 azonal) and hence the following, all with a series,: *Vaccinium gaultherioides* (zonal), *Loiseleuria procumbens* (intrazonal), *Rhododendron ferrugineum* (zonal), *Erica carnea* (zonal), *Rhododendron hirsutum* (zonal), *Juniperus nana* (intrazonal), *Alnus viridis* (intrazonal), *Salix helvetica* (intrazonal), *Salix arbuscula* (intrazonal), *Salix waldsteiniana* (intrazonal), *Myricaria germanica* (azonal), *Hippophaë rhamnoides* (azonal), *Salix incana* (azonal).

The series characterized by trees are the following: *Pinus sylvestris* 10 series (1 zonal, 1 extrazonal, 4 intrazonal, 4 azonal), *Abies alba* 8 series (intrazonal), *Fagus sylvatica* 6 (zonal), *Alnus glutinosa* 6 (azonal), *Picea abies* 6 (5 zonal, 1 azonal), *Fraxinus excelsior* 4 (2 zonal, 2 azonal), *Pinus cembra* 3 (zonal), *Larix decidua* 3 (zonal), *Acer pseudoplatanus* 3 (intrazonal), *Ostrya carpinifolia* and *Fraxinus ornus* 3 (zonal), *Alnus incana* 1 (azonal), *Carpinus betulus* 1 series (zonal), *Quercus petraea* 1 (zonal), *Quercus pubescens* 1 (zonal), *Taxus baccata* 1 (intrazonal), *Quercus ilex* 1 (extrazonal), *Salix alba* 1 (azonal), *Salix pentandra* 2 (azonal), *Populus alba* 1 (azonal), *Betula pubescens* 1 (azonal), *Quercus robur* 1 (azonal).

First of all, it must be observed that some species only participate in the construction of zonal series, such as *Fagus sylvatica*, *Picea abies* (which also participates in an extrazonal series), *Pinus cembra*, *Larix decidua*, *Ostrya carpinifolia*, *Fraxinus ornus*, *Quercus petraea*, *Quercus pubescens*, *Carpinus betulus*. A second group of species participates only in intrazonal series: *Abies alba*, *Acer pseudoplatanus*, and *Taxus baccata*. A third group participates in the extrazonal series, with two species, *Quercus ilex* and *Picea abies*. A fourth group participates in the azonal series, with *Alnus glutinosa*, *Alnus incana*, *Salix alba*, *Salix eleagnos*, *Salix purpurea*, *Salix cinerea*, *Salix pentandra*, *Populus alba*, and

Betula pubescens. Three species enter into the composition of series of varied typology, namely *Pinus sylvestris*, which can be zonal, intrazonal, extrazonal and azonal, *Fraxinus excelsior*, which is zonal and azonal, and *Picea abies* which can be zonal and extrazonal.

Other arboreal species (such as *Acer platanoides*, *Tilia europaea*, *Tilia platyphyllos* and others) are present in various vegetation series, characterized by other species. For example, *Acer platanoides* is found in the *Calamintho grandiflorae-Aceri pseudoplatani* sigmetum and in some series of *Fagus sylvatica*, *Tilia europaea*e *Tilia platyphyllos* in *Galio laevigati-Carpinetum betuli*, etc.

The species that characterizes the highest number of vegetation series is *Pinus sylvestris,* 10 series found from the steppe-like slopes of the Val Venosta to the Prealps. Another species with broad ecological adaptation, albeit limited on the slopes of the montane belt with northern exposure, is *Abies alba*, with 8 vegetation series, while *Fagus sylvatica* is part of 6 vegetation series. *Alnus glutinosa* as well is part of a notable number of vegetation series (6, to be precise, all azonal), in both riparian and swampy series. In contrast, the two conifer species that are typical of the alpic and endoalpic sectors, namely *Picea abies* and *Pinus cembra*, are part of only 5 and 3 series, respectively. This confirms the specialization of these two species in the phytogeographic and ecological sense. The same can be said of *Larix decidua*, which characterizes only 3 vegetation series; the widespread presence of this species today is due to anthropic factors that promoted its spread. In the hilly belt, *Carpinus betulus* characterizes only one vegetation series, as does *Quercus petraea*, which however also rises to the lower montane belt. *Ostrya carpinifolia* and *Fraxinus ornus* both characterize 3 series.

The vegetation series are thus distributed in the various altitudinal belts: nival 5, alpine 18, subalpine 21, montane 28, hill 11, and azonal 26. The highest number of series is found in the montane belt, with 28 vegetation series; the montane belt presents the broadest altitudinal range, from 1000 to 1800 m, and the greatest variety of environments, which favour the development of many intrazonal series that contribute to raise the number of series. Regarding the type of substratum, excluding the azonal series, there are 41 vegetation series on siliceous and 40 on calcareous substrata.

3.7 The Zonality of Vegetation Series

The zonal series are those that give the general impression to the plant landscape. The extrazonal series join the previous ones locally. The intrazonal series appear in limited areas within spaces occupied by the zonal series. The azonal series interrupt the former, according to the geomorphology. The zonality of the vegetation series is evident in the maps of the distribution of the series that were prepared in reference to the UTM network (Venanzoni 1990), a few examples of which will be presented below.

3.7.1 Zonal Series

Zonal series develop in average climatic and pedologic conditions according to the latitude (vegetation zones) and the altitude (vegetation belts). The chain of these series from the hilly belt to the nival one is the hallmark of the vegetation landscape of the Alps of the Trentino-Alto Adige Region, as presented by Pedrotti and Gafta (2003), referring, however, to the megageoseries. The three prealpic, alpic and inner-alpic sectors are easily distinguishable, even though a notable block of vegetation series repeat in them, for example *Fraxino orni-Ostryeto carpinifoliae* (Fig. 3.7a), *Luzulo niveae-Querceto petraeae*, *Adenostylo glabrae-Abieteto albae*, *Adenostylo glabrae-Piceeto abietis*, *Larici-Piceeto* sigmetum (Fig. 3.8), *Erico-Pinion mugo*, *Vaccinio-Pineto mugo*, *Rhododendro-Vaccinieto myrtilli*, *Seslerion albicantis*, *Caricion curvulae* (Fig. 3.9) and a few others.

Other vegetation series are common to two sectors, for example prealpic and alpic (beech woods, including *Carici albae-Fageto sylvaticae* sigmetum) (Fig. 3.10a) or alpic and inner-alpic (for instance, *Arabidi-Querceto pubescentis* sigmetum).

Some vegetation series are present exclusively in one sector; the prealpic sector is characterized by hornbeam groves (*Galio laevigati-Carpineto betuli* sigmetum) (Fig. 3.11) and by some beech woods (*Cardamino pentaphylli-Fageto* and *Calamagrostio villosae—Fageto* sigmetum); the alpic sector hosts only one exclusive vegetation series, the *Carici albae-Abieteto albae* sigmetum; in the inner-alpic sector the beech woods are substituted by mountain spruce forests (*Luzulo-Piceeto abietis* sigmetum), and there are also xeric pine woods of *Astragalo vesicarii-Pineto sylvestris* sigmetum.

3.7.2 Extrazonal Series

The only distinct extrazonal series in the Alps of Trentino-Alto Adige are the *Celtidi—Querceto ilicis*and the *Vaccinio vitis-ideae—Pineto sylvestris*. In both cases, it is held that the forest formations in which they grow (mediteranean ilex groves and continental pine woods) form distinct vegetational zones on a primarily climatic basis in the Mediterranean region and in the central-european and north-east-european provinces of the Eurosiberian region, respectively. In fact, these two series are, respectively, northern and southern disjunctions of zonal Mediterranean ilex groves and continental north-east-european pine woods. Notwithstanding the main climatic determinism of these disjunctions, the presence of these two series in the territory under study is also caused by some edaphic conditions, that is, the calcareous substrate for the ilex groves and strongly leached soils for the pine woods.

In regard to the associations of zonal ilex groves and pine woods (of the Mediterranean and of the north-east European), the extrazonal ones found in Trentino-Alto Adige include particular species due to the influx of alpic elements such as *Erica herbacea*, *Luzula nivea*, *Rhododendron ferrugineum* and *Amelanchier ovalis*, etc., respectively.

While the *Celtidi australis-Quercetum ilicis* series is distributed only in the low Lagarina Valley and in the Sarca Valley (Fig. 3.12), the *Vaccinio vitis-idaeae—Pineto sylvestris* series is limited only to the continental sector of the region, north of Trento.

3.7.3 Intrazonal Series

The intrazonal series appear in one or two altitudinal belts near each other in particular environmental conditions, due to exposure, declivity and topography (Ivan 1979).

All the vegetation series of the small nival valleys should be considered intrazonal in the alpine belt because they occupy shady, concave-shaped habitats, where the snow remains much longer than it does in surrounding habitats.

The *Loiseleuria* series forms an edaphic-xerophilous site climax inasmuch as it tends to occupy the windy, well-drained summits; thus this series is intrazonal in the belt of the dwarf shrubs. The *Erica* series is diffused in discontinuous islands because it is linked to steep habitats with very superficial soil, which thus are marked by temporary hydric deficit. This series is thus intrazonal in the belt of the dwarf and contorted shrubs. The dwarf juniper series (*Juniperus nana*) tends to grow on the sunny, and thus relatively more xeric slopes, and hence is intrazonal in the belt of the contorted shrubs (*Rhododendro-Vaccinion*). The Swiss mountain pine (with sorb) is limited to the sites with debasified soils, developed, however, on calcareous-dolomitic substrates. This series is thus intrazonal inasmuch as it establishes itself in the belt of the basophilus Swiss mountain pine woods (*Rhododendro hirsuti-Pineto* sigmetum). In the same belt, in meso-hygrophilous sites (gorges, shady slopes or those with hillside springs), the Swiss mountain pine woods mentioned above are substituted by willow groves of the intrazonal series of *Saliceto caesio-arbusculae* sigmetum.

The sparse woods of larch and cembran pine (*Larici-Pineto cembrae* sigmetum) are substituted on the sunny, more xeric slopes by almost monospecific woods formed of larch and with ground cover dominated by dwarf juniper. This series (*Junipero Lariceto* sigmetum) should thus be considered intrazonal.

The edapho-hygrophilous series of spruce (*Listero-Piceeto* sigmetum), forms enclaves within the zonal series of *Larici-Piceeto* sigmetum in small valleys and upland plains. Similarly, the series of spruce on moist screes (*Asplenio-Piceeto*) appears here and there within the series of *Luzulo-Piceeto* sigmetum.

All the fir woods form a site climax in the belt of the beech woods and spruce forests, as they tend to appear in the third median of the shady slopes (Gafta 1994). For instance, the *Calamagrostio villosae-Abieteto* sigmetum is intrazonal in the belt of the subalpine spruce forest (*Homogyno-Piceeto* sigmetum) or the subalpine beech wood (*Calamagrostio villosae-Fageto* sigmetum), the *Cardamino pentaphylli-Abieteto* sigmetum in the belt of *Cardamino pentaphylli-Fageto* sigmetum, the *Carici albae-Abietetum* (Fig. 3.10b) in the belt of *Carici albae-*

Fagetum, and so on. However, the presence of fir-woods cannot be explained only in terms of the ecological conditions, but also by taking into consideration their postglacial contraction into refuge areas.

The rocky series of beech woods (*Taxo-Fageto* sigmetum) is a particular case, found in the belt of the *Carici albae-Fageto* sigmetum, on rocky walls and slopes.

Only in the suboceanic sector, in moist sites, the montane beech woods (*Cardamino pentaphylli-Fageto* sigmetum) are substituted by maple groves that form the series head of the *Calamintho grandiflorae-Acereto pseudoplatani* sigmetum.

Most of the Scotch pine series should be considered intrazonal inasmuch as the series heads associations establish themselves precisely in the zonal belts of other series. These pine woods form a site climax due to particular geopedological conditions, that set a more xeric character than that of the other nearby zonal sites, *Erico-Pineto sylvestris* occupies rocky and semi-rocky habitats in the belt of the *Carici albae-Fagetum* in the suboceanic zone, or of the *Adenostylo glabrae-Piceeto* in the (sub)continental one; *Molinio litoralis-Pineto sylvestris* occupies habitats with outcroppings of marly limestones in the belt of the*Carici albae-Fageto* in the suboceanic zone, or of the *Adenostylo glabrae-Piceeto* in the (sub) continental zone; *Chamaecytiso purpurei-Pineto sylvestris* (Fig. 3.7b) establishes itself on the rocky slopes of the belt of *Fraxino orni-Ostryeto carpinifoliae*, exclusively in the suboceanic zone; *Antherico liliaginis-Pineto sylvestris* is found on eroded and sunny cliffs in the belt of *Luzulo niveae-Querceto petraeae*, above all in the (sub)continental zone.

Of the four series listed, the most ecologically dissonant, in terms of the zonal series in which it is established, is *Antherico liliaginis-Pineto sylvestris*, which has the most xerophilous character. In fact, it should be noted that the presence of intrazonal pine woods and fir woods cannot be explained solely on the basis of ecological conditions; rather, an explanation should also take into consideration postglacial contractions of these forest formations to areas of refuge.

3.7.4 Azonal Series

The azonal series found can be divided into three categories on the basis of the edaphic characteriestics created by a stagnant and/or phreatic hydric regime: riparian series, palustrine series and series of peaty substrates. These vegetation series characterize above all the plant landscape of valley bottoms and high plains.

The lower course of the wider valleys that cross the hilly belt is characterized by *Populeto albae*sigmetum; in the montain belt one observes the presence of *Carici remotae-Fraxineto* sigmetum and sometimes (only in the alpic sector) *Pruno-Fraxineto* sigmetum.

The palustrine series are only present in sites with flat morphology (lacustrine basins, etc.) and are formed of the following sigmeta: *Saliceto pentandrae*, *Alno-Saliceto pentandrae*, *Molinio coeruleae-Pineto sylvestris* sigmetum.

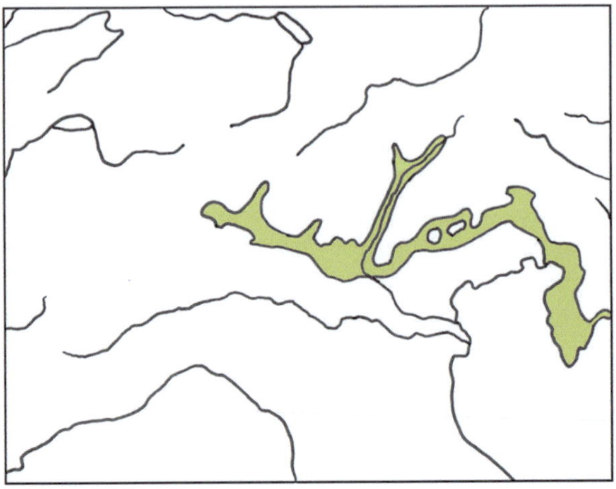

Fig. 3.13 Distribution of associations of alliance of *Orno-Ostryon* (from Bohn et al. 2000 modified)

The peaty series are conditioned by substrates with the presence of peat: *Vaccinio uliginosi-Pinetum sylvestris*, *Vaccinio uliginosi-Betuleto pubescentis*, *Sphagno girgensohnii-Piceeto* and *Pino mugo-Sphagneto* sigmetum.

Today in the Trentino-Alto Adige Region, the distribution of these vegetation series is very patchy, of a residual type, because of the reclamations conducted over the centuries.

The vegetation landscape of the Trentino-Alto Adige Region is characterized by 128 vegetation series. This high number is the consequence of multiple causes, such as latitudinal extension, lithological substrates, geomorphological conditions and phytoclimate. It demonstrates the complexity and the great phytogeographic diversity of the territory under study, with a very marked gradient, passing from the prealpic to the endoalpic sector, which can be highlighted by referring to both the flora (including the endemisms) and the vegetation series and megageoseries.

3.8 Concluding Remarks

Comparing the southern side of the central Alps, from the Brenner Pass to the Padanian plain, and the northern side, from the Brenner Pass to the Bavarian plain, one notes similarities and differences in the altitudinal sequence and in the typology of the vegetation.

The Padanian plain is markedly different from the Bavarian one; in the former the potential vegetation is forests of *Carpinion betuli* s.l., while in the latter, it is formed by lowland forests of *Fagion sylvaticae*.

In the hilly belt, the southern side is characterized by the *Orno-Ostryon* alliance (Fig. 3.13), which is missing in the northern side where *Fagion sylvaticae* alliance grows in the hilly belt (Fig. 3.14). As on the plains, so also in the hilly belt there is a dissymmetry between the two sides, as can be noted on Ozenda's vegetation map of the alpine arc (1984) and on Bohn et al's vegetation map of Europe (2000).

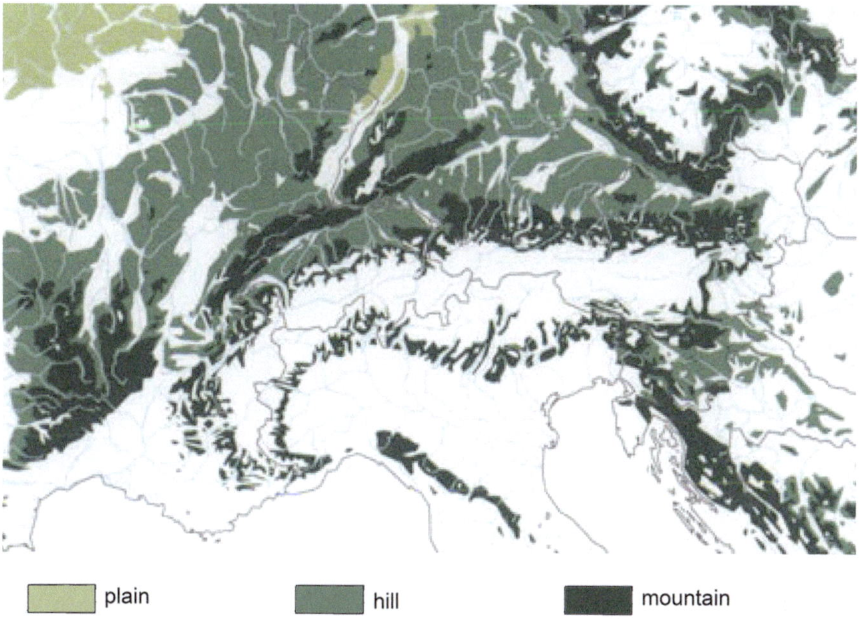

Fig. 3.14 Distribution of the associations of the *Fagion sylvaticae* alliance in the alpine chain and surrounding zones; in the Italian side of the Alps *Fagus sylvatica* is distributed only in the montane belt; in the northern side it is found in the lowland, hilly and montane belts (kindly supplied by Udo Bohn—Bad Godesberg)

In the montane, subalpine and alpine belts, the alliances are the same on both sides (*Fagion sylvaticae*, *Erico-Pinion*, *Dicrano-Pinion*, *Abieti-Piceion*, *Piceion excelsae*, *Rhododendro-Vaccinion*, *Caricion curvulae*, *Seslerion coeruleae,* and *Caricion firmae*, etc., Fig. 3.15); in this case, therefore, there is symmetry as regards the alliances, though often the associations are not the same.

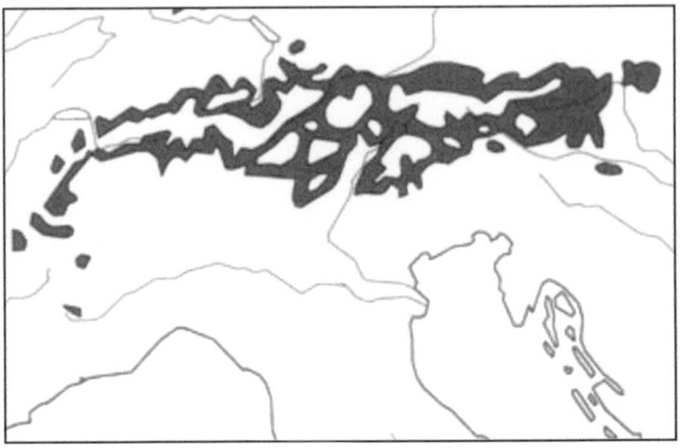

Fig. 3.15 Distribution of the *Rhododendro-Vaccinion* and *Piceion excelsae* alliances in the alpine chain (*dark color*) and of the *Caricion curvulae* and *Rhododendro-Vaccinion* alliances (*white colored areas* inside the *dark colored area*)

References

Bohn U, Gollub G, Hettwer C (2000) Karte der natürlichen Vegetation Europas. Massstab 1: 250.000. Legende. Karte. Bundesamt f. Naturschutz, Bad Godesberg

Boiti I, Lasen C, Saffaro Boiti T (1989) La vegetazione della Val Venegia. Manfrini, Calliano (Trento)

Braun-Blanquet J (1946) Über die Pflanzendecke der Gebirge zwischen Zernez und Reschenscheideck. Communication SIGMA 91:1–23

Braun-Blanquet J (1961) Die inneralpine Trockenvegetation. Fischer, Stuttgart

Erschbamer B (1992) Zwei neue Gesellschaften mit Krummseggen (Carex curvula ssp. rosae, Carex curvula ssp. Curvula) aus den Alpen – ein Beitrag zur Klärung eines alten ökologischen Rätsels. Phytocoenologia 21(1–2):91–116

Gafta D (1994) Tipologia, sinecologia e sincorologia delle abetine nelle Alpi del Trentino. Braun-Blanquetia 12:1–69

Gafta D, Pedrotti F (1994) Phytosociological and ecological research in a protected areas as basis for its management: the example of Loppio lake (North Italy). Applied Veg Ecol. In: Proceedings of 35th symposium IAVS Shangai East China. Normal University Press, pp 31–40

Gafta D, Pedrotti F (1998) Fitoclima del Trentino – Alto Adige. Studi Trent Sci Nat 73:55–111

Ivan D (1979) Fitocenologie şi vegetaţia Republicii socialiste romăna. Editura Didactica, Bukarest

Kiem J (1992) Ein Tamariskenvorkommen in Sarntal. Ber Bayer Bot Ges 63:139–143

Mayer H (1977) Karte der natürlichen Wälder des Ostalpenraumes. Centralbl. Gesamte Forstwesen 94:147–153

Minghetti P (1996) Analisi fitosociologica delle pinete a Pinus mugo del Trentino (Italia). Doc Phytosoc 16:461–503

Minghetti P (2003) Le pinete a Pinus sylvestris del Trentino-Alto Adge (Alpi Italiane): tipologia ecologia e corologia. Braun-Blanquetia 33:1–95

Ozenda P (1984) La végétation de l'arc alpin. Council of Europe, Strasbourg

Ozenda P (1985) La végétation de la chaîne alpine dans l'espace montagnard européen. Masson, Paris

Ozenda P (2013) May an biogeographical pattern be transferred into other major mountain ranges? Coll Phytosoc XXIX:467–473

Pedrotti (2013) Plant and vegetation mapping. Springer, Heidelberg

Pedrotti F (1965–1968) Carta fitosociologica della vegetazione della media Val di Sole (Trento). Società Geografica, Firenze

Pedrotti F (1981) Carta della vegetazione del Foglio Trento. Coll Progr Fina. Promozione qualità ambiente. CNR AQ/1/17: 1–38
Pedrotti F (1982) Carta della vegetazione del Foglio Mezzolombardo. Coll Progr Final. Promozione qualità ambiente. CNR AQ/1/85: 1–25
Pedrotti F (1992) La vegetazione. In: Ambiente M (ed) Relazione sullo stato dell'Ambiente. Ist. Poligr. Stato, Roma, pp 94–100
Pedrotti F (1993) Saggio di carta della vegetazione della Regione Trentino-Alto Adige (scala 1: 500.000). Boll Ass Ital Cart 87(88):149–154
Pedrotti F (1997) Geobotanik und Landschaftskartierung – Beispiele aus Italien. Ber Reinh-Tüxen Ges 9:123–127
Pedrotti F (2004) Ricerche geobotaniche al Laghestel di Piné (1967-2001). Braun-Blanquetia 35:1–55
Pedrotti F (2006) Vegetation series of the Trentino-Alto Adige region (Italy). Polish Botanical Studies 22:427–435
Pedrotti F (2010) Le serie di vegetazione della Regione Trentino-Alto Adige. In: Blasi C (ed) La vegetazione d'Italia. Palombi, Rome, pp 83–109
Pedrotti F (2015) Chorology and phytosociology of sessile oak [Quercus petraea (Mattuschka) Liebl.] in Trentino-Alto Adige (East-Central-Alps) of Northern Italy. In: Box EP, Fujiwara K (eds) Warm-temperate deciduous forests around the northern hemisphere. Springer, Heidelberg, pp 175–180
Pedrotti F (in print) La certographie géobotanique en Europe. Documents phytosociologiques, III série
Pedrotti F, Gafta D (1996) Ecologia delle foreste ripariali e paludose dell'Italia. L'uomo e l'ambiente 23:1–165
Pedrotti F, Gafta D (2003) Phytogeographical approach to distinction of vegetation megageoseries in the Alps of Trentino-Alto Adige (with map 1: 250.000). Report CEA Trento 30, pp 1–18
Pedrotti F, Orsomando E, Cortini Pedrotti C (1974) Carta della vegetazione del Parco Nazionale dello Stelvio (Notizia esplicativa). Amm. Parco Naz. Stelvio, Bormio
Pedrotti F, Gafta D, Minghetti P (2003) Map of natural potential vegetation. In: Pedrotti F, Gafta D (eds) Phytogeographical approach to distinction of vegetation megageoseries in the Alps of Trentino-Alto Adige (with map 1: 250.000). Report CEA Trento 30, pp 1–18
Peer T (1995) La vegetazione naturale dell'Alto Adige. Provincia autonoma di Bolzano, Bolzano
Pitschmann H, Reisigl H, Schiechtl HM, Stern R (1980) Karte der aktuellen Vegetation von Tyrol 1/100.000 – VII Teil, Blatt 10, Ötztaler Alpen-Meran. Doc Cart Écol 23:47–68
Prosser F (1997) La vegetazione del Monte Altissimo di Nago (Monte Baldo). Atti Acc Roveretana Agiati 247(VIIB):115–178
Prosser F, Sarzo A (2003) Il Monte Brione: aspetti vegetazionali, indicizzazione del pregio botanico ed esigenze conservazionistiche (Riva del Garda, Trentino, Italia settentrionale). Atti Acc Roveretana Agiati IIIB:149–195
Reinalter R (2003) Zur Flora der Sedimentgebiete in Umkreis der südrätischen Alpen, Livignasco, Bormiese und Engiadin'Ota (Schweiz–Italien). Birkhauser
Rivas Martìnez S (1987) Memoria del mapa de series de vegetaciòn de Espana. ICONA, Madrid
Rivas Martìnez S (1996) Bioclimatic map of Europe (scale 1: 10.000.000). Cartographic Service. University of Leon, Leon
Schiechtl HM, Stern R (1976) Karte der aktuellen Vegetation von Tyrol 1/100.000 – VI Teil, Blatt 11, Brixen-Pustertal. Doc Cart Écol 17:73–84
Schiechtl HM, Stern R, Zoller H (1982) Karte der aktuellen Vegetation von Tyrol 1/100.000 – VIII Teil, Blatt 9, Silvretta-Engadin-Vinschgau. Doc Cart Écol 25:67–88
Theurillat JP (1992) Étude et cartographie du paysage végétal (symphytocoenologie) dans la région d'Aletsch (Valais, Suisse). Matér Levé Géobot Suisse 68:1–384
Venanzoni R (1990) La presenza di Carex appropinquata Schumacher in Trentino- Alto Adige. Inform Bot Ital 22(3):194–196

Bryophytes and Mesoclimatic Gradients Along a Transect of the Adige Valley (Central Alps)

4

Michele Aleffi

Abstract

Present work identify the relations between the distribution of some species of bryophytes and the principal mesoclimatic gradients along the north-south section of the Adige Valley between the border of Veneto—Trentino Alto Adige and the entrance to the Venosta Valley. The correlations between the distribution of some groups of species of bryophytes belonging to different chorotypes and the main climatic variables of the transect make it possible to formulate a few biogeographic hypotheses.

Keywords

Bryophytes • Mesoclimatic gradients • Adige Valley (Central Alps)

4.1 Introduction

The mountain chain of the Alps, because of its morphological and geological complexity and its extension, which determines a great variety of mesoclimates, is marked by a high degree of bryophyte diversity. Even though alpine regions cover only 30% of Italy's surface, the specific richness in bryophytes of this area constitutes 93.7% of Italy's entire patrimony of bryophytes.

The goal of the present work was to identify the relations between the distribution of some species of bryophytes and the principal mesoclimatic gradients along the north-south section of the Adige Valley between the border of Veneto-Trentino (45°42′ latitude north) and the entrance to the Venosta Valley (46°40′ latitude north). This choice was motivated first of all by the state of bryological and phytoclimatic knowledge (Aleffi et al. 2008; Gafta and Pedrotti 1998) about the

M. Aleffi (✉)
University of Camerino, Via Pontoni, 5, 62032 Camerino, MC, Italy
e-mail: michele.aleffi@unicam.it

F. Pedrotti (ed.), *Climate Gradients and Biodiversity in Mountains of Italy*, Geobotany Studies, https://doi.org/10.1007/978-3-319-67967-9_4

Fig. 4.1 Map of the Trentino-Alto Adige Region with indication of the course of the Adige River and of the meteorological stations along the valley bottom

territory of Trentino compared to the rest of the Italian alpine crescent, and by the relatively high number of meteorological stations that have been recording temperature and precipitation data for a significantly long time (Fig. 4.1).

4.2 Materials and Methods

In order to highlight some climatic characteristics such as continentality and humidity, the following bioclimatic indexes were calculated: the intra-annual extreme temperature range, the index of hygric continentality (Gams 1931; Michalet 1991; Pache et al. 1996), the index of pluvial continentality (Angot 1906; Daget 1977) and the compensated ombrothermic index (Rivas Martínez 1994).

Forty of the numerous bryophyte species reported along the transect of the Adige Valley were chosen because their ecological behaviour in terms of distribution corresponds their limited ecological range (stenoecia). Statistics were generated from the data with multivariate analysis methods using the cluster and ordination approaches (Guiot 1986; Hubalek and Borakova 1988; Retuerto and Carballeira 1991; Podani 1995). To this end, both continuous climatic data and discontinuous chorological data in the form of binary values were used.

4.3 Discussion of the Results

The classification of the meteorological stations on the basis of the matrix of the averages of the maximum and minimum monthly temperatures is presented in the form of dendrogram in Fig. 4.2; it indicates two distinct sections of the valley. The first, from Ala to Trento, is characterized by a milder climate, above all in the winter, while the second, from Lavis to Merano, presents more marked extremes of temperature. The same analysis, done on the basis of the averages of monthly precipitation (Fig. 4.3), reveals a different division of the transect: the first part, from Ala to Salorno, has greater rainfall and an evident secondary pluviometric maximum in November (the padano-pre alpine regime), while the second, from Bolzano to Merano, has lower annual rainfall and a continental type annual pluviometric regime.

Supplementary information can be obtained from the flowchart analysis of the meteorological stations that result from the matrix of pluviothermic data (Fig. 4.4). The transect can be divided into three relatively homogeneous sections; however, the greatest discontinuity is seen between Salorno and Bolzano, demonstrating that

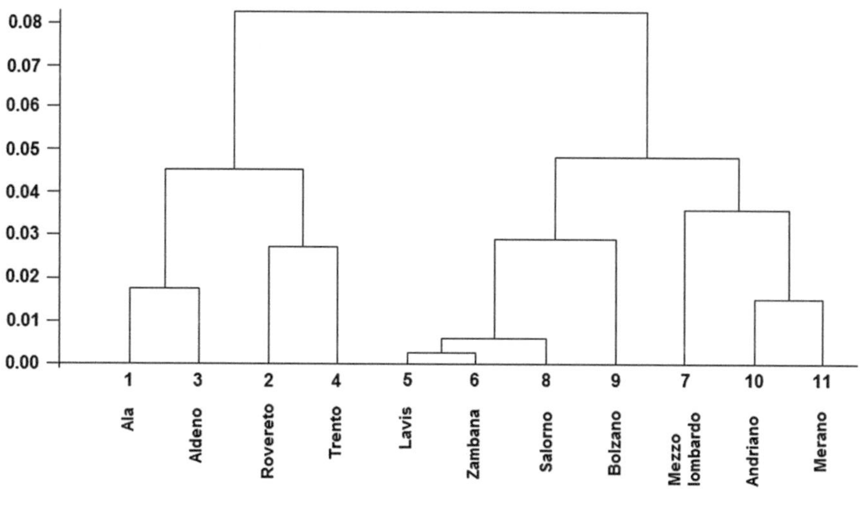

Fig. 4.2 Dendrogram of the meteorological stations obtained on the basis of the average values of the maximum and minimum monthly temperatures

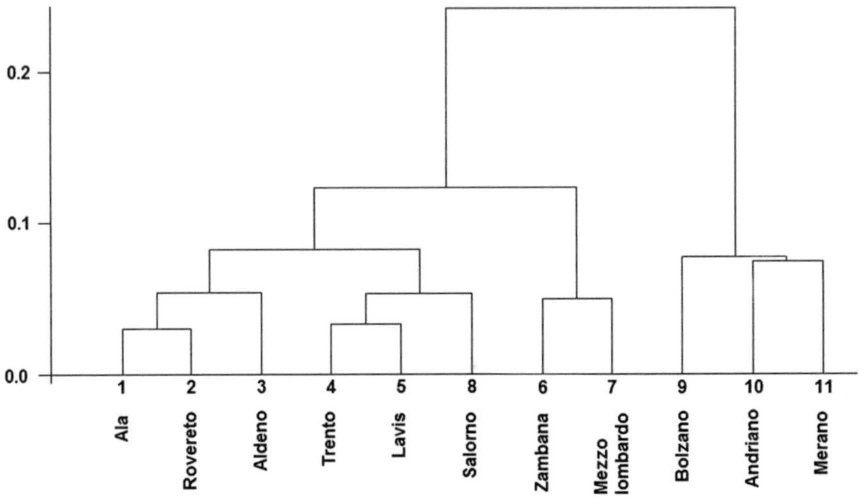

Fig. 4.3 Dendrogram of the meteorological stations obtained on the basis of the average values of the monthly precipitation

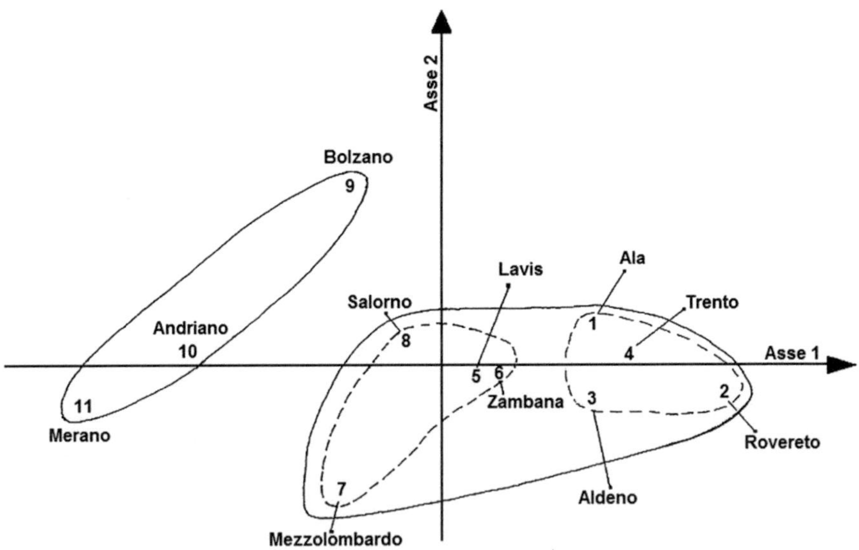

Fig. 4.4 Flowchart of the meteorological stations obtained through analysis of the main components applied to the matrix of the temperature and pluviometric data. Axis 1 and 2 explain respectively 56% and 17% of the total variance. Axis 1 is well correlated with the minimum temperatures of February (0.964), December (0.942) and January (0.938), as well as with the precipitation of January (0.874). Axis 2 is well correlated with the maximum temperatures of February (0.800), October (0.799) and, in part, with the precipitation of November (−0.573)

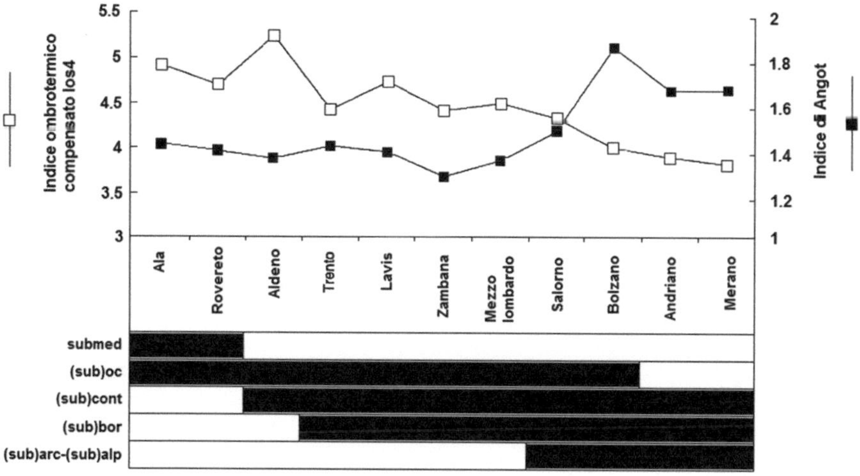

Fig. 4.5 Trend of the compensated ombrothermic index and of the Angot index and interval of the presence of the bryophyte species, grouped by chorotypes, along the transect

the pluviometry is more discriminating than the temperature. Axis 1 is well correlated with the averages of the minimum temperatures of the winter months and with the average rainfall of the month of January. Consequently, the layout of the stations along that axis shows the stations of the lower section (as far as Trento) opposite those located above Salorno. Axis 2 is quite well correlated with the maximum temperatures of the month of February and with the average precipitation of the month of November; the most continental station, Bolzano, remains separate. While these results agree with the values of the bioclimatic indices, the variation of these indices along the transect reveals two main trends: the increase of pluvial (Angot) and hygric (Gams) continentality and the reduction of humidity (Rivas Martínez) from south to north (Figs. 4.5 and 4.6).

Based on these graphics of the presence of the species, grouped by chorotypes (Table 4.1), and of the trend of the main climatic variables along the transect, the following considerations can be made (Marchesoni 1946–1947; Aleffi et al. 1997):

- The (sub)mediterranean species are present only in the lower section of the valley, below Aldeno, because of the milder winter climate from the point of view of temperature;
- The (sub)oceanic species occupy most of the transect, except the section above Bolzano, where the climate is decidedly more continental in terms of temperature and moisture;
- The (sub)continental species occupy a complementary section to that occupied by (sub)mediterranean species, characterized by a more accentuated overall continentality;
- The (sub)boreal species are present in the upper section, above Aldeno, with considerably lower winter minimum temperatures;

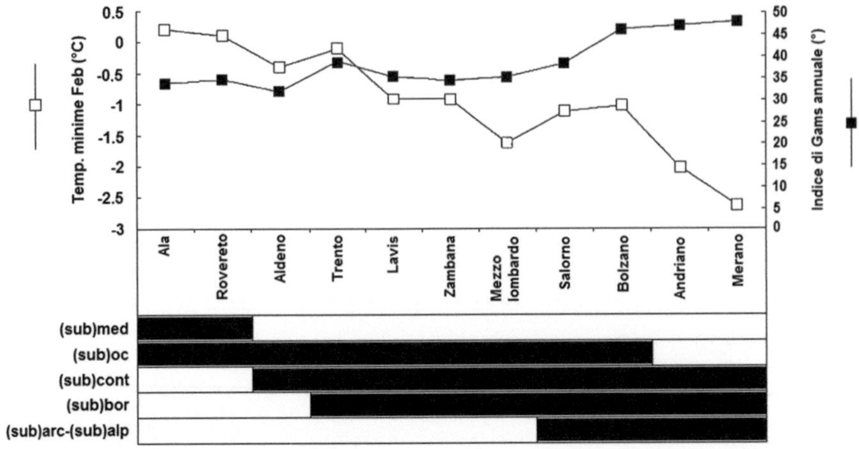

Fig. 4.6 Trend of the averages of the minimum temperatures of February and of the annual index of Gams and interval of presences of bryophyte species, grouped by chorotype, along the transect analyzed

– The (sub)arctic-(sub)alpine species best indicate the great climatic discontinuity above Salorno.

4.4 Conclusions

The correlations between the distribution of some groups of species belonging to different chorotypes and the main climatic variables of the transect make it possible to formulate a few biogeographic hypotheses:

– The montane sector of the Adige Valley has provided an immigration route for submediterranean and suboceanic species and at the same time a postglacial refuge for the arctic-alpine species that emigrated toward the north of the continent;
– The upper section of the valley bottom hosts a high number of continental and boreal species because it is subjected to intense temperature inversions.

Table 4.1 Presence of the main indicator species for climate, grouped by chorological elements along the transect

Meteorological Stations	ALA	ROVERETO	ALDENO	TRENTO	LAVIS	ZAMBANA	M.LOMBARDO	SALORNO	BOLZANO	ANDRIANO	MERANO
submed											
Tortella nitida	+	+	.	.	.	.	.	.	.	.	.
(sub)oc											
Barbula crocea	+	.	.	+	.	.	.	+	.	.	.
Bryum dichotomum	+	.	.	+	.	.	.	.	+	.	.
Campylopus pilifer	.	.	.	.	.	.	.	+	+	.	.
Paraleucobryum sauteri	+	.	.	.	.	.	.	.	.	.	.
Rhynchostegiella teesdalei	+	.	.	.	.	.	.	.	.	.	.
Sphaerocarpos michelii	.	.	.	+	.	.	.	.	.	.	.
(sub)cont											
Brachythecium laetum	.	.	+	+	+	.	.	.	+	+	+
Mnium spinulosum	.	.	.	+	+	+	.	.	.	.	.
Breidleria pratensis	.	.	.	.	.	.	.	.	+	.	.
Asterella saccata	.	.	.	.	.	.	.	.	.	.	+
Frullania jackii	.	.	.	.	.	.	.	.	.	.	+
Notothylas orbicularis	.	.	.	.	.	.	.	.	.	.	+
Callicladium haldanianum	.	.	.	.	.	.	.	.	.	.	+
(sub)bor											
Dicranum muehlenbeckii	.	.	.	+	.	.	.	.	+	+	+
Tortula cernua	.	.	.	+	+	.	.	.	.	.	+
Barbilophozia attenuata	.	.	.	.	.	.	.	.	+	+	.
Bazzania tricrenata	.	.	.	.	.	.	.	.	+	+	.
Ptilidium ciliare	.	.	.	.	.	.	.	.	+	+	.
Ptilidium pulcherrimum	.	.	.	.	.	.	.	.	+	+	.
Cynodontium strumiferum	.	.	.	.	.	.	.	.	+	+	.
Hygrohypnum ochraceum	.	.	.	.	.	.	.	.	+	.	.
Polytrichastrum pallidisetum	.	.	.	.	.	.	.	.	+	.	.
Racomitrium fasciculare	.	.	.	.	.	.	.	.	+	.	.
Anastrophyllum saxicola	.	.	.	.	.	.	.	.	.	+	.
Lophozia incisa	.	.	.	.	.	.	.	.	.	.	+
Didymodon icmadophilus	.	.	.	.	.	.	.	.	.	.	+
(sub)arc-(sub)alp											
Encalypta affinis	.	.	.	.	.	.	.	+	+	+	+
Cnestrum schisti	.	.	.	.	.	.	.	.	+	+	+
Gymnomitrion corallioides	.	.	.	.	.	.	.	.	+	+	+
Stegonia latifolia	.	.	.	.	.	.	.	+	+	.	.
Dicranodontium asperulum	.	.	.	.	.	.	.	.	+	+	.
Tayloria splachnoides	.	.	.	.	.	.	.	.	+	+	.
Marsupella brevissima	.	.	.	.	.	.	.	.	+	.	.
Encalypta microstoma	.	.	.	.	.	.	.	.	+	.	.
Didymodon giganteus	.	.	.	.	.	.	.	.	.	.	+
Hygrohypnum smithii	.	.	.	.	.	.	.	.	.	.	+
Andreaea frigida	.	.	.	.	.	.	.	.	.	.	+
Jungermannia confertissima	.	.	.	.	.	.	.	.	.	.	+
Campylopus schimperi	.	.	.	.	.	.	.	.	.	.	+

References

Aleffi M, Cortini Pedrotti C, Gafta D (1997) Considerazioni briogeografiche mediante un'analisi bioclimatica lungo un tratto della Val d'Adige. Rev Valdôtaine Histoire Naturelle 51:379–396

Aleffi M, Tacchi R, Cortini Pedrotti C (2008) Check-list of the hornworts, liverworts and mosses of Italy. Bocconea 22:1–255

Angot A (1906) Etude sur le régime pluviométrique de la méditerranée. C R Congr Soc Sav:120–134

Daget P (1977) Le bioclimat méditerranéen: Caractères generaux, modes de caractérisation. Vegetatio 34(1):1–20

Gafta D, Pedrotti F (1998) Fitoclima del Trentino-Alto Adige. Studi Trent Sci Nat, Acta Biol 73 (1996):55–111

Gams H (1931) Die klimatische Begrenzung von Pflanzenarealen und die Verteilung der hygrischen Kontinentalität in den Alpen. Zeitschr Ges Erdkunde Berlin:321–346

Guiot J (1986) Sur la détermination de régions climatiques quasi-homogènes. Revue de Statistiques Appliquées 34(2):15–34

Hubalek Z, Borakova M (1988) Evaluation of climatic similarity between areas in biogeography. J Biogeogr 15:409–418

Marchesoni V (1946-1947) Analisi fitogeografica degli elementi floristici del bacino atesino. Mem Mus St Nat Ven Tridentina VII:3–78; VIII:3–40

Michalet R (1991) Nouvelle synthèse bioclimatique des milieux méditerranéens. Application au Maroc septentrional. Rev Ecol Alp, tome I:60–80

Pache G, Aime S, Michalet R (1996) A simple model for the study of the altitudinal rainfall gradient, applied in the tyrolian orographic complex. Rev Ecol Alp 3:13–20

Podani J (1995) Sin-Tax 5.02 Mac. Computer Programs for Multivariate Data Analysis on the Macintosh system. Scientia Publishing, Budapest

Retuerto R, Carballeira A (1991) Defining phytoclimatic units in Galicia, Spain, by means of multivariate methods. J Veg Sci 2:699–710

Rivas Martínez S (1994) Clasificacion bioclimatica de la Tierra. Folia Botanica Matritensis 12:1–20

Are Non-native Plant Species a Threat to the Alps? Insights and Perspectives

5

Consolata Siniscalco and Elena Barni

Abstract

Mountain systems throughout the world are recognized as important biodiversity hot spots with rich and distinct flora, compared to the surrounding lowland regions and provide ecosystem services that are important for agriculture and forestry, for touristic, recreational and cultural activities. For these reasons the spread of invasive non-native species can be a threat to the mountains, increasing direct and indirect costs or decreasing the income of the inhabitants.

Several studies have been carried out in the Alps, in order to highlight the distribution of alien species and to elucidate the factors that influence their occurrence, considering climate, disturbance and, in particular, human impact and introduction pathways as extrinsic factors while the plant traits of alien species, propagule pressure, and the invasibility of habitats have been considered as intrinsic factors driving the arrival and success of the new comers. All results highlight that the invasion process in the Alps is at the beginning and it is still possible to limit its impacts, since the invasions at high elevations will occur over long periods of time so that researchers and managers have the opportunity of responding in time to the threat.

In this work results that have been obtained from research carried out in the Western Italian Alps are analyzed and discussed considering similar results obtained in other mountain regions. The following topics are discussed: (a) the distribution of alien species along the elevation gradient; (b) the influence of climatic factors on the distribution; (c) species invasive capacity; (d) habitat invasibility; (e) distribution models.

C. Siniscalco (✉) • E. Barni
Department of Life Sciences and Systems Biology, University of Torino, Viale Mattioli, 25-10125 Torino, Italy
e-mail: consolata.siniscalco@unito.it

F. Pedrotti (ed.), *Climate Gradients and Biodiversity in Mountains of Italy*,
Geobotany Studies, https://doi.org/10.1007/978-3-319-67967-9_5

Keywords
Alien species • Alps • Mountain regions • Invasion • Temperature

5.1 Introduction

Mountain systems throughout the world are recognized as important biodiversity hot spots with rich and distinct flora, compared to the surrounding lowland regions (Körner 2002), since they offer variability of climatic and soil conditions over short distances and very often host not only rare and endemic species of biogeographical interest, but also rare plant communities and unique ecosystems. Besides their importance as biodiversity hot spots, mountains also provide ecosystem services that are important for agriculture and forestry, such as grazing activities and the related food production or the provision of wood, as well as for touristic, recreational and cultural activities. For these reasons, the spread of alien species in mountain regions can not only threaten the biodiversity of native species biodiversity but can also influence the economy and human health.

Mountain regions, at least at high elevations, were considered immune to the spread of invasive alien species until the end of the last century for many reasons: they are generally less invaded than the lowland regions and the number of alien species is generally consistently lower (Pauchard et al. 2009; McDougall et al. 2011a, b).

However, studies conducted in the last few years have demonstrated that mountains are increasingly threatened by invasive alien plants as a consequence of the introduction of their propagules and human impact (Johnston and Pickering 2001; Bear et al. 2006).

Owing to their recent invasion, mountain systems provide the opportunity of studying the process from the beginning, and offer good opportunities to analyze mechanisms and causes of invasion over short spatial distances along the elevation gradient (Becker et al. 2005; Körner 2007; Kueffer et al. 2008). This opportunity also shows the possible impacts that non-native species could have in the future, considering the on-going climatic and land use changes.

Several studies have been carried out in order to highlight the distribution of alien species and to elucidate the factors that influence their occurrence, considering climate, disturbance and, in particular, human impact and introduction pathways as extrinsic factors while the plant traits of alien species, propagule pressure, and the invasibility of habitats have been considered as intrinsic factors driving the arrival and success of the new comers.

The comparisons of the occurrence and invasion patterns of alien species in mountain regions throughout the five continents, have mainly been carried out by the MIREN (Mountain Invasion Research Network) group, and they have shown that although there are differences in the mountain regions, general trends can be established.

Areas at highest elevation show consistently low invasion levels (Pauchard et al. 2009) although a recent comparison of different mountain systems throughout the world has shown that the altitudinal richness of non-native species varies from region to region but with a unimodal distribution in sub-tropical oceanic mountains (Tassin and Riviere 2003; Arévalo et al. 2005) and a continuous decrease in temperate mountains (Pauchard and Alaback 2004; Becker et al. 2005; Mallen-Cooper and Pickering 2008), with areas above the timberline not being affected by non-natives (Becker et al. 2005; McDougall et al. 2005).

As far as the plant invasion mechanism is concerned, four main factors have been suggested to explain the general rule of decreasing numbers of non-native species along the elevation gradient (Pauchard et al. 2009): the first is the general lack of pre-adaptation of non-native plants to the harsh environmental conditions, the second is the decreasing levels of anthropogenic disturbance, the third is the low invasibility of native plant communities, and the fourth is the low number of introduced plant propagules.

Almost all of the introduced non-native species grow at low elevations in their native ranges (Arévalo et al. 2005; Daehler 2005), and are thus thought to be unable to survive the effects of frost and snow and the short growing season (Rejmánek 1989; Rejmánek and Randall 1994). Disturbance is a well-documented factor that plays an important role in increasing the opportunities for alien invasion, by reducing the resistance of native plant communities and increasing the availability of sites and resources (Pauchard and Alaback 2004; Parks et al. 2005). For this reason, increased tourist activity at higher elevations is of concern (Pickering and Hill 2007), particularly in the European mountains.

Mountains below which there are areas with dense populations, intense agriculture and urbanization, are probably more subject to invasion, due to the high level of traffic (Gelbard and Belnap 2003; McDougall et al. 2005), than isolated mountains where propagules are less likely to be introduced (Stohlgren et al. 1999; Pauchard and Alaback 2004).

Alien species have only recently begun to spread in the European Alps, at medium and high elevation, and very few species have been detected above the timberline. Until now, their occurrence is particularly frequent in human influenced habitats, mainly road verges and riversides, which stretch along valleys and act as corridors for the spread of alien species spread (Becker et al. 2005; Pickering and Hill 2007; Alexander et al. 2009).

Results that have been obtained from research carried out in the Western Italian Alps are analyzed in this work considering the above mentioned factors and are discussed considering similar results obtained in other regions. The following topics are discussed: (a) the distribution of alien species along the elevation gradient; (b) the influence of climatic factors on the distribution; (c) species invasive capacity; (d) habitat invasibility; (e) distribution models.

5.2 Distribution of Alien Species Along the Elevation Gradient

In the Western Italian Alps non-native species occurrence decreased remarkably with increasing elevation (Siniscalco et al. 2011, Fig. 5.1) as detected in many other mountain regions (Arévalo et al. 2005; Becker et al. 2005; Marini et al. 2009;

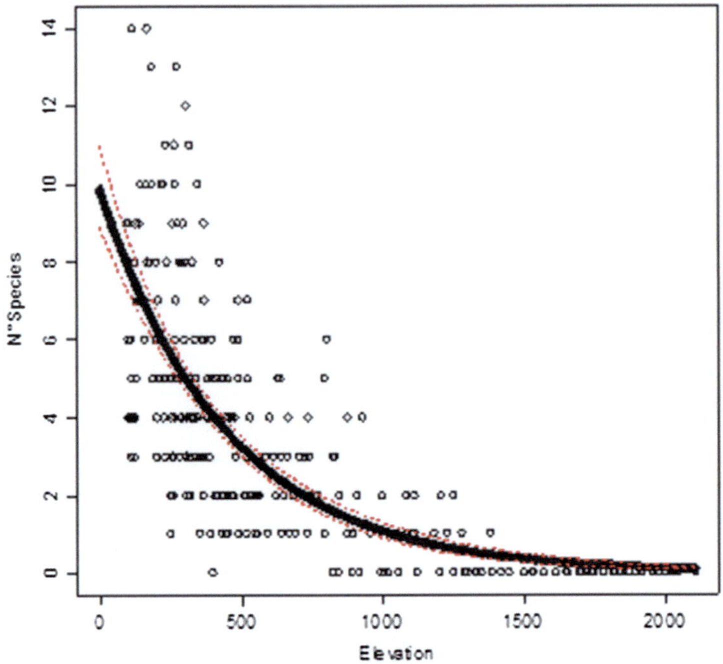

Fig. 5.1 Estimated relationship for non-native plant species richness with respect to the elevation. *Dotted lines* represent 95% C.Is. (from Siniscalco et al. 2011)

Pauchard et al. 2009; Alexander et al. 2011), thus confirming that elevation is a key factor in limiting plant invasion. The results were obtained using a systematic sampling design with 250 plots regularly spaced 2 km from each other, from 100 to 2100 m a.s.l., along river courses and in valleys with different climates and land uses.

All the 68 non-native species found in the sampling plots were detected at low elevation, between 100 and 450 m, (with a very high variability among plots), while only 29 species were found from 400 to 800 m and only 10 and 4 were found between 800 and 1150 m and between 1150 and 1500 m (Fig. 5.2). No alien species were found in the sampling plots above this elevation. This distribution confirms that species propagules come from the lowland and spread into the valleys along their axis, transported by man along the main roads or rivers, during soil or sand transport. Most of the species are invasive in Italy (71%), while 25% are naturalized and only 4% are casual (status assigned according to Celesti-Grapow et al. 2009). As far as life forms are concerned, it emerged that there was a prevalence of annual species (56%) while there was a prevalence of herbaceous species (28% of the total) among the perennials (44%).

The geomorphological characteristics of the valleys in the Western Alps, with high mountain ridges, very few mountain passes between valleys and only a few roads from the lowlands, explain why, until now, all the species have been

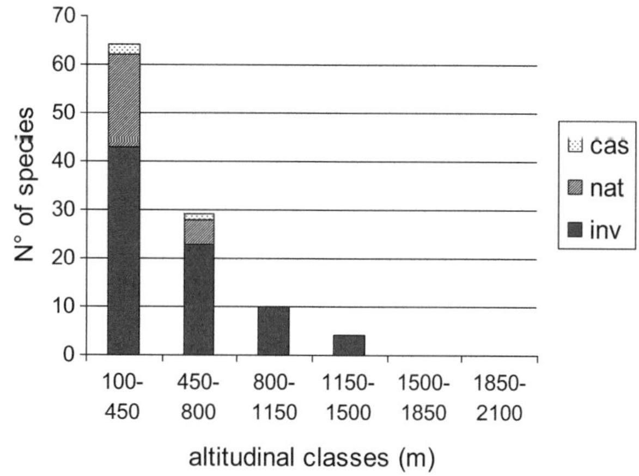

Fig. 5.2 Number of casual, naturalized and invasive species in the elevation classes (from Siniscalco et al. 2011)

introduced from the plain to the mountain regions. The lowland alien flora closeby is the main determinant of the composition of the mountain region's alien species, as was also concluded in an extensive research in which the alien floras of 13 mountain regions in the world was compared (McDougall et al. 2011a). In the present case, the arrival of the species from the Po Plain, which is rich in alien species, can be supported by their distribution pattern as well as by their life forms and status, which are typical of the lowland species occurring in the adjacent Po Plain region.

The progressive reduction in human activities with increasing elevation explains the lower number of propagule introductions at higher elevations. Since very few species have been introduced deliberately above the timberline, with the exception of the alpine gardens and for ski piste restorations, the transport of seeds and propagules has only occurred in valleys with a road and a tunnel or a pass connecting two adjacent valleys. A different situation has been detected in the Swiss Alps (Becker et al. 2005), where roads arrive at relatively high elevations and alien species have been detected at higher elevations than in the Western Italian Alps. This confirms the great importance of human impact in determining the arrival of non-native species in high elevation mountain regions.

5.3 Climatic Factors Influencing Distribution

The reasons for the non-native species decrease with elevation have been related to the climate filtering hypothesis (Weiher and Keddy 1995), which can explain part of the distribution pattern. Recently, the decreasing occurrence of non-native species along the elevation gradient has been explained by the directional ecological filtering (Alexander et al. 2011) which states that species with a broad climatic tolerance can survive difficult climatic conditions and that all the others are progressively suppressed, filtered along the elevation gradient.

Once elevation has been chosen as a predictor of invasive species richness, it has to be considered as a proxy of many implicit factors that could, separately or in combination, limit the spread of alien plants.

The relationships between climate and biological invasions have been extensively investigated to predict the potential distribution of alien plants using a climate matching approach (Broennimann et al. 2007; Richardson and Thuiller 2007). Some criticisms have been made about correlative approaches by pointing out that (i) the observed species distributions may be constrained by non-climatic processes (Guisan and Zimmermann 2000; Pearson and Dawson 2003) such as disturbance, biotic interactions and limited propagule dispersal, (ii) the present limits may not be in equilibrium with the environment because the range may still be expanding (Collingham et al. 2000) and (iii) correlative approaches do not reveal the mechanisms by which climate drives plant distribution (Williamson 2006).

Despite these limitations, species distribution models based on statistical relationships between occurrence data and underlying climatic conditions are increasingly used to describe or predict the spatial patterns of biological invasions (Thuiller et al. 2005): their results can generate rigorous hypotheses on the causes of the distribution patterns observed, and provide useful knowledge to steer preventive and control measures. Moreover, it is extremely important to identify the climatic constraints that affect the spread of alien species in the general global-warming scenario (Bear et al. 2006; Walther et al. 2009).

In the Western Italian Alps we analyzed the role of climatic constraints in shaping the distribution of alien plant species along rivers from 100 to 2100 m a.s.l. (Barni et al. 2012). Climate variables were calculated from the data recorded by 145 meteorological stations and interpolated by a multiple regression approach. Both the richness and the occurrence of aliens were modelled. In particular, the relationships between the occurrence of alien plants and (1) elevation or (2) the climatic variables, were tested by applying generalized linear models and generalized linear mixed models; the model parameters obtained were used to estimate the upper elevation limits of alien occurrence and their related climate values. A steep decrease in alien species richness with elevation was found, with the probability of alien species occurrence decreasing by half for each 100 m increase in elevation. Minimal adequate models based on (1) non-transformed climatic variables and (2) derived PCA values, confirmed that the occurrence of alien plant species along the elevation gradient was positively related to the minimum temperature, the mean temperature and the heat sum for the spring season, rather than to the incidence of absolute minimum temperature and frost days, as is usually assumed. Considering the climatic models that were developed, we estimated the relationship between alien plant occurrence and the most explicative predictor, which was the minimum spring temperature. This relationship showed a probability of alien occurrence lower than 10% when the minimum spring temperature threshold of 0.1 °C was exceeded (Fig. 5.3) (at an elevation of 1475 m), a probability lower than 5% when −0.5 °C was exceeded (1578 m) and a probability lower than 1% when −1.9 °C was exceeded (1834 m).

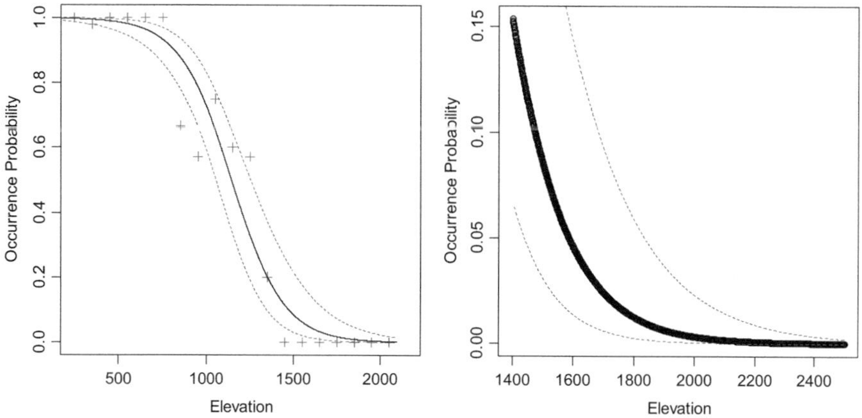

Fig. 5.3 Plots of the estimated logit function for the predictive model explaining alien species occurrence with respect to elevation gradient. (**a**) Presence/absence of alien species in each sampled area plotted against elevation; *black crosses* represent the mean occurrence values calculated within 100 m wide elevation classes; *dotted lines* represent the estimated confidence intervals (95%) for the logit function. (**b**) Detail of the probability function estimated for alien plant species occurrence with respect to the elevation range between 1400 and 2500 m. *Dotted lines* represent 95% CI (from Barni et al. 2012)

The minimal adequate model that we used confirmed that the distribution of alien plant species along rivers in the studied alpine valleys was determined to a great extent by climate, which explained the largest amount of variation (76%), while the effects of human pressure, edaphic and biotic factors or other climatic variables were less important, and corresponded to the residual variation that had not been explained.

This result can be related to the relatively homogeneous disturbance level occurring all along the elevation gradient in the river habitats. Conversely, studies on extensive areas, characterized by very different disturbance levels, have pinpointed the difficulties in discriminating the effects of the elevation gradient from human impact (Nogués-Bravo et al. 2008) or highlighted the importance of human impact in determining the alien distribution. An extensive research based on two different databases on native and alien species in the Bergamo, Brescia and Trento provinces, has highlighted that both human pressures and temperature selection are important processes that underpin the observed filtering (Marini et al. 2013). The relationship between species richness and temperature for natives was remarkably dependent on life-form, while the absence of life-form-dependent responses for alien species supports the hypothesis that the distribution of alien plant species richness was more closely related to propagule pressure and disturbance.

These results show that the process of invasion is still at the beginning in the studied areas in the Alps and that the introduced species are only a small selection of the invasive species occurring in the lowlands.

5.4 Species Invasive Capacity

Several researches have tried to establish which plant functional traits make an alien species a successful invader. Among the considered traits, the production of a high number of seeds, a high growth rate and a high biomass production are the most important, but the production of allelopathic compounds, as well as the symbiosis with nitrogen fixing bacteria, or with mycorrhizal fungi, can also enhance the competitive capacity of the species. The specific composition of the invaded communities is consequently influenced by the growth of the most invasive non-native species that alter the plant and soil communities and ecosystem processes reducing the native biodiversity (Manchester and Bullock 2000; Hejda et al. 2009 and references therein).

Most of the species that occur in the Alps are invasive species adapted to the lowlands and they are characterized by high ecological plasticity and tolerance to the mountain climatic conditions (as suggested by Becker et al. 2005, Fig. 5.4). One example is the East Asian *Fallopia japonica* (Houtt.) Ronse Decraene, one of the 100 world's worst invasive alien species (Lowe et al. 2000), that has a strong and developed rhizome system, that accumulates reserves which have been transferred during autumn from the leaves and stems. The rhizomes survive below the soil during the winter and allow the plant to produce a huge number of new stems and leaves during the early spring outcompeting the native species and reducing biodiversity (Mincheva et al. 2014). The species has been observed at 1580 m in the Western Italian Alps (Siniscalco et al. 2011; Bouvet et al. 2013).

Similarly, *Heracleum mantegazzianum* Sommier & Levier, an invasive species native to the Caucasus mountains is characterized by similar traits: it is a biennial/perennial species with a huge taproot which accumulates reserves and allows the plant to produce large leaves very rapidly during spring, immediately after snowmelt (Curtaz et al. 2011). During autumn, the plant effectively transfers metabolites to the taproot and during winter it does not have any living part aboveground, and is thus able to tolerate extremely low air temperatures and survive under the snow. The species has been observed in the Aosta Valley at 1500 m a.s.l. where it produces viable seeds in spite of a relatively short growing season and survives in Alpine Botanical Gardens at 2000–2200 m a.s.l. This species produces severe contact dermatitis and is thus a threat to human health and also to biodiversity. *Fallopia japonica* and *Heracleum mantegazzianum* have divergent plant traits compared to most of the native species, in line with the "try harder" hypothesis (Tecco et al. 2010; Dainese and Bragazza 2012) and successfully outcompete the natives, altering the ecosystem functioning.

Unlike these two species, other species such as *Papaver croceum* Ledeb. (native to the Altay regions in Cental Asia), *Arnica nevadensis* A. Gray (mountain regions of NW America), *Geranium cinereum* Cav. (S e W Balcans, C and S Italy and the Pyrenees) and *Achillea lingulata* Wadst. & Kit. (Carpathians) (Fig. 5.5), which are pre-adapted to high mountain conditions, have similar traits to the native congeners and can consequently spread in alpine pastures and debris after their direct introduction into Alpine Gardens, at high elevations. *Papaver croceum* Ledeb. has been

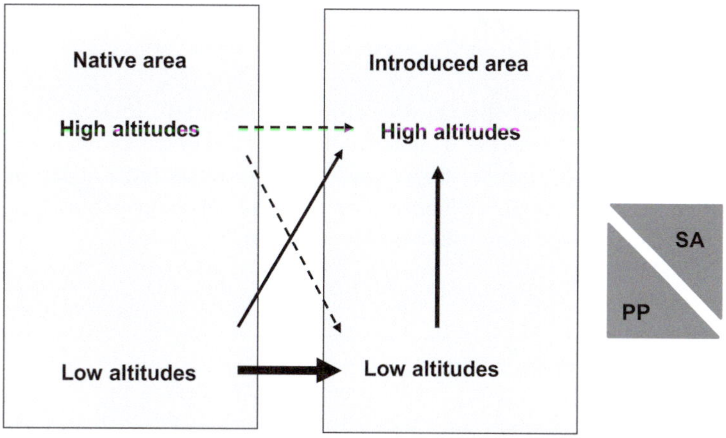

Fig. 5.4 Conceptual diagram on probable routes by which alien species enter an high-mountain region (relative importance of the routes indicated by *arrow thickness*). Because the principal transport routes connect lowland areas, most introduced populations may lack adaptations to higher mountain areas. Therefore, specific adaptation (SA) is likely to increase in importance with altitude as a constraint for invasion. Conversely, propagule pressure (PP) is likely to decrease in importance relative to environmental effects and partly as a result of increasing distance from large source populations in the lowland (from Becker et al. 2005 re-drawn)

Fig. 5.5 *Geranium cinereum* Cav., *Arnica nevadensis* A. Gray, and *Achillea lingulata* Wadst. & Kit., three alien species cultivated at the Alpine Botanical Garden Chanousia at Col du Petit St. Bernard

observed in alpine habitats in Switzerland as a pioneer species in debris and disturbed sites, having similar ecology to the native *Papaver aurantiacum* Loisel. (Irl et al. 2013). The other three species are cultivated in the Alpine Botanical Garden Chanousia at Col du Petit St. Bernard (2150 m) at the border between Italy and France, produce a high number of seeds and tend to spread around and outside the garden. A careful control is needed every summer to avoid their spreading.

These species are in line with the "join the local" hypothesis (Tecco et al. 2010; Dainese and Bragazza 2012) as they present functional convergence with the native species.

The occurrence of alien species with different plant traits, that can tolerate the harsh climatic conditions of the mountains in different ways, once again highlights that the invasion process is at the beginning and that its evolution will depend on the species that are introduced, deliberately or not, and on the availability of novel niches created by human activities (Marini et al. 2011).

5.5 Habitat Invasibility

Most of researches on alien species in mountain systems have been carried out along roads and rivers, since the occurrence of these species is still limited to disturbed habitats. In the Italian Alps, native and non-native species occurrence has been established in 19 habitat types on the basis of data of Flora Alpina (Aeschimann et al. 2004), the most complete flora available for the Italian alpine region (Dainese and Bragazza 2012). Based on 107 non-native species, it was established that the habitats with the highest proportion of alien species were river beds and banks (about 19%), ruderal vegetation (about 17%), freshwater habitats (9%), megaforbs (5%), and hay meadows (3%). The proportion of alien species was negligible in dry grasslands and wet meadows, (<1%). Forests, heathlands, peatlands, and herbaceous alpine habitats (e.g., alpine grasslands, snow beds, screes and rocky slopes) were not colonized by alien species.

The results of this work show that the invasion success is linked to human or natural disturbance regimes in association with high nutrient supply and warm climate (Becker et al. 2005; Walter et al. 2005; Marini et al. 2009). On the contrary, environmentally stressful conditions, such as low temperature, pronounced drought, low nutrient availability, and lower propagule pressure, due to the distance from human activities (Pyšek et al. 2002; Chytry et al. 2008), have until now protected the montane and alpine zone habitats of the Alps from alien plant invasion.

Species at low and medium elevations, below 1600 m a.s.l. in the Western Alps, in line with the "try harder hypothesis", are usually ruderal or ruderal-competitive species sensu Grime (1979) and can tolerate harsh climatic conditions thanks to their ecological plasticity. They can also be subjected to environmental selection of the genotypes which are probably less able to invade disturbed habitats. Species at higher elevations, in line with the "join the local" hypothesis, which are stress-tolerant species sensu Grime, have to be introduced at high elevations and could, in the future, become a real threat, not only for the biodiversity of the disturbed habitats but also of pastures and other natural or semi natural high altitude habitats.

5.6 Distribution Models

Species distribution models (SDMs) have been used to project the potential distribution of non-native species and to support effective mitigation or eradication activities in the adventive range at a global or, more often, regional scale (Guisan and Thuiller 2005; Wilson et al. 2007; Gallien et al. 2010). By fitting a relationship between the occurrence of the observed species and environmental factors, SDMs can be used to assess the role of natural or anthropogenic disturbance, vegetation types, and climatic variables as invasion drivers. Finally, SDMs can provide "early warnings" in areas of major concern for the potential spread of invasive species, and help evaluate whether the species is in the introduction, colonization, or establishment stage of invasion (Theoharides and Dukes 2007). Dispersal constraints must be included in SDMs for species in the first two stages of invasion, in order to optimize their fit by taking into account the geographical limitations in the seed source (Vaclavik and Meentemeyer 2009).

We have recently monitored the occurrence of *Senecio inaequidens* DC. in the Aosta Valley Region (Vacchiano et al. 2013) and used SDMs in order to model its distribution with the following aims: (i) to evaluate whether the species is in the introduction, colonization, or establishment stage of invasion; (ii) to detect the environmental factors that drive the invasion process; and (iii) to highlight the potential range of distribution, in order to support control activities by either private citizens or public administrations. *Senecio inaequidens* is one of the fastest spreading non-native species in Europe, mainly in mountain systems. Like congeneric species, it contains pyrrholizidine alkaloids that are toxic to livestock and to human beings through milk consumption. It is pre-adapted to mountain climatic conditions, since it is native to the mountain regions of South Africa, from 1400 to 2850 m a.s.l. *S. inaequidens* displayed a very fast spread in the Aosta Valley since 1990 and occurs all along the axes of the main and of some secondary valleys (Fig. 5.6), where it has invaded not only disturbed habitats, but also dry pastures with low vegetation covers.

The modeling framework was a stepwise generalized linear model (GLM), which used gridded presence/absence data and environmental predictors such as topography, climate, land use, and anthropogenic and natural disturbances. GLMs were fit both with and without an additional independent variable, to take into account the current dispersal limitations.

The species was positively associated with roads and rivers, southern slopes, and negatively to elevation. However, it was found at an elevation of 1600 m, thus showing the ability to reach higher elevations than those observed for other invasive alien species, and confirming to be pre-adapted to mountain conditions. The difference between the species distribution models, with and without dispersal constraints (Figs. 5.7 and 5.8), suggested that the availability of the seed sources still limits the potential distribution of the species, rather than the environmental variables, and that the realized regional niche differs to a great extent from the equilibrium niche. When the limitations to the seed source cease (i.e., in the establishment stage), the species will likely invade large areas that are currently

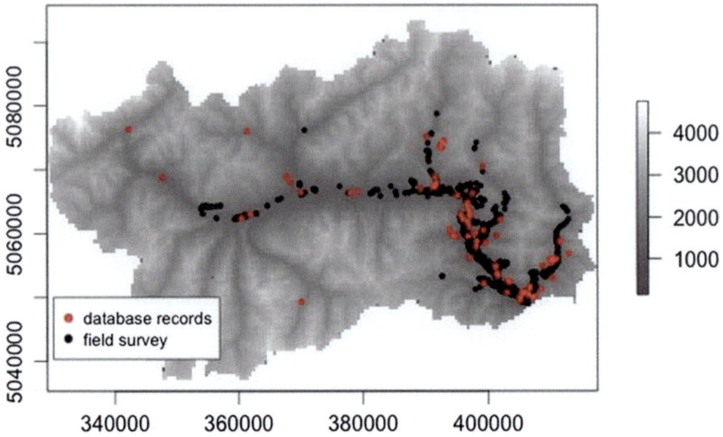

Fig. 5.6 Digital elevation (m) model of the study area (coordinate system: UTM WGS 1984, zone 32N) and presence points for *S. inaequidens* from the regional flora database (*red*, 1985–2012) and from the intensive field survey (*black*, 2009–2010) (from Vacchiano et al. 2013)

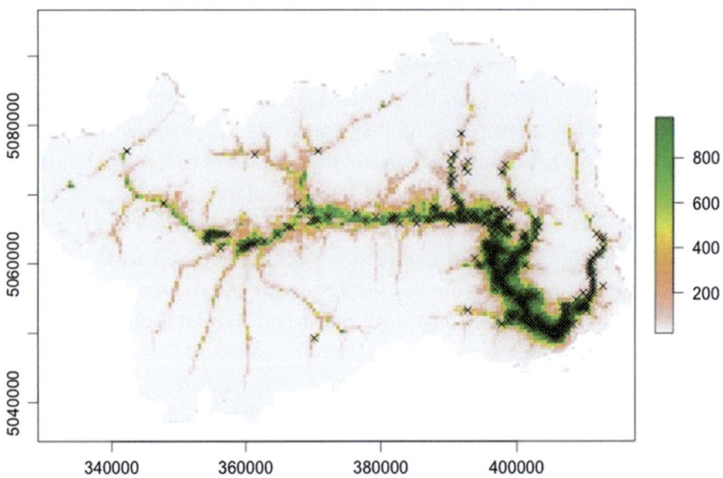

Fig. 5.7 Presence probability (×1000) of *S. inaequidens* in the study area according to a distribution model that includes dispersal constraints. The *crosses* indicate actual presence points (from Vacchiano et al. 2013)

characterized by pastures and grasslands with native species of high agricultural importance. The invasion of *S. inaequidens* should therefore be considered a serious threat, due to its potential to invade mountain regions, and in particular to colonize habitats used for grazing and forage, thus leading to a high risk for cattle and human health.

Since the grasslands on the south facing slopes in the mapped area can be particularly invasible by the studied species, awareness campaigns have already

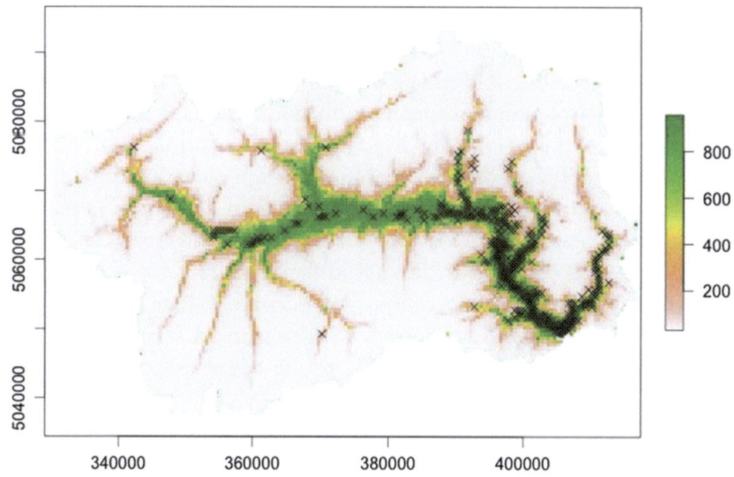

Fig. 5.8 Presence probability (×1000) of *S. inaequidens* in the study area according to a distribution model that does not include dispersal constraints. The *crosses* indicate actual presence points (from Vacchiano et al. 2013)

been carried out in the region (Curtaz et al. 2011). It is highly recommended that these campaigns are continued for farmers, professionals, and garden center owners in order to disseminate knowledge on this species and its risks in the Aosta Valley. The inclusion of *S. inaequidens* in the "black list" of the regional law for the conservation of alpine flora (L.R. 45/2009) will help to transfer information on this species and support eradication activities, particularly at medium and high elevations.

This example can explain the possible applications of species distribution models to understand whether a non-native species is in the introduction, colonization or establishing stage of invasion and which environmental factors are driving the invasion process, as well as to highlight the potential future range of distribution.

A critical point is that genetic changes, due to the selection of the best adapted ecotypes, are not considered in the models, and this increases the uncertainty of the potential range of distribution.

5.7 Conclusions and Perspectives

The results obtained in the researches in the Western Italian Alps and in other Alpine Chain regions indicate that areas at high altitudes are not especially prone to alien invasion, due to temperature constraints, low propagule pressure, low disturbance and low invasibility of the majority of montane and alpine habitats, a result that is in agreement with what has been detected in other mountain systems (Pauchard et al. 2009). However, some particularly invasive species, presenting some favourable traits ("try harder" hypothesis) which are absent in the native species, are able to invade medium and high elevation disturbed ecosystems. Future adaptation of these species could enhance the invasion of less disturbed habitats. On the other hand, the introduction of species with similar traits to the native congeners ("join the locals" hypothesis), which have rarely occurred until now, for the cultivation or restoration of disturbed habitats at high elevation, could be a threat to the biodiversity of rare and unique alpine habitats in the future. In particular the cultivation of pre-adapted species coming from mountain systems of the world in Alpine Gardens has to be monitored to avoid their spread.

Since these invasions at high elevations will occur over long periods of time, due to the slow production of seeds in short growing seasons, researchers and managers have the opportunity of responding in time to the threat associated with invasive species (McDougall et al. 2011a, b).

Future projections on the spread of non-native species in the Alps, in the case of climate warming, are possible and will help managers face the threat; however, projections never consider the probable genetic changes that the invasive species will encounter. Moreover, while the newly introduced species will undergo genetic changes, also the native species, plants, animals, bacteria and fungi, will also undergo genetic variation due to the introduction of the alien species. Finally, some enemies (fungi or insects) from the native range of the newcomers, could be introduced deliberately or not in the Alps and help control the spread of the species.

In this framework, preventing the arrival of new non-native species (McDougall et al. 2011a, b) is undoubtedly the most effective action to avoid impacts on biodiversity, economy and human health in the Alps. The recent regional laws in some Italian alpine regions as Valle d'Aosta (L.R. 45/2009) and Lombardy (L.R. 10/2008) consider the importance of dissemination of the possible impacts of some non-native species, in particular to nurserymen who grow ornamental species and to those technicians who are involved in the management and restoration of disturbed areas.

The European Parliament has recently proposed a legislation that will be discussed in the next months to stem the spread of invasive alien species banning the species declared of "Union concern". This new legislation highlights that, as in the Alps, the alien species do not stop at the national borders and thus more and better coordinated action by member states are needed to tackle the threat.

References

Aeschimann D, Lauber K, Moser DM, Theurillat JP (2004) Flora alpina. Zanichelli, Bologna

Alexander JM, Edwards PJ, Poll M, Parks CG, Dietz H (2009) Establishment of parallel altitudinal clines in traits of native and introduced forbs. Ecology 90:612–622

Alexander JM, Kueffer C, Daehler CC, Edwards PJ, Pauchard A, Seipel T, MIREN Consortium (2011) Assembly of nonnative floras along elevational gradients explained by directional ecological filtering. PNAS 108:656–661

Arévalo RJ, Delgado JD, Otto R, Naranjo A, Salas M, Fernández-Palacios JM (2005) Distribution of alien vs. native plant species in roadside communities along an altitudinal gradient in Tenerife and Gran Canaria (Canary Islands). Perspect Plant Ecol 7:185–202
Barni E, Bacaro G, Rocchini D, Falzoi S, Spanna F, Siniscalco C (2012) Distribution of non-native plant species along the elevation gradient beside rivers located in the Alps: establishing climatic constraints. Plant Ecol 213:757–767
Bear R, Hill W, Pickering CM (2006) Distribution and diversity of exotic plant species in montane to alpine areas of Kosciuszko National Park. Cunninghamia 9:559–570
Becker T, Dietz H, Billeter R, Bushmann H, Edwards PJ (2005) Altitudinal distribution of alien plant species in the Swiss Alps. Perspect Plant Ecol 7:173–183
Bouvet D et al (eds) (2013) Piante esotiche invasive in Piemonte. Riconoscimento, distribuzione, impatti. Regione Piemonte, Torino
Broennimann O, Treier UA, Müller-Schärer H, Thuiller W, Peterson AT, Guisan A (2007) Evidence of climatic niche shift during biological invasion. Ecol Lett 10:701–709
Celesti-Grapow L, Alessandrini A, Arrigoni PV, Banfi E, Bovio M, Brundu G et al (2009) The inventory of the alien flora of Italy. Plant Biosyst 143:386–430
Chytry M, Maskell LC, Pino J, Pyšek P, Vilà M, Font X, Smart SM (2008) Habitat invasions by alien plants: a quantitative comparison among Mediterranean, subcontinental and oceanic regions of Europe. J Appl Ecol 45:448–458
Collingham YC, Wadsworth RA, Huntley B, Hulme PE (2000) Predicting the spatial distribution of non-indigenous riparian weeds: issues of spatial scale and extent. J Appl Ecol 37:13–27
Curtaz A, Talichet M, Barni E, Bassignana M, Masante D, Pauthenet Y, Siniscalco C (2011) Specie esotiche invasive e dannose nei prati dimontagna. Institut Agricole Régional, Aosta
Daehler CC (2005) Upper-montane plant invasions in the Hawaiian Islands: patterns and opportunities. Perspect Plant Ecol 7:203–216
Dainese M, Bragazza L (2012) Plant traits across different habitats of the Italian Alps: a comparative analysis between native and alien species. Alp Bot 122:11–21
Gallien L, Münkemüller T, Albert CH, Boulangeat I, Thuiller W (2010) Predicting potential distributions of invasive species: where to go from here? Divers Distrib 16:331–342
Gelbard JL, Belnap J (2003) Roads as conduits for exotic plant invasions in a semiarid landscape. Conserv Biol 17:420–432
Grime JP (1979) Plant strategies, vegetation processes, and ecosystem properties. Wiley, Chichester
Guisan A, Thuiller W (2005) Predicting species distribution: offering more than simple habitat models. Ecol Lett 8:993–1009
Guisan A, Zimmermann NE (2000) Predictive habitat distribution models in ecology. Ecol Model 135:147–186
Hejda M, Pyšek P, Jarosyk V (2009) Impact of invasive plants on the species richness, diversity and composiion of invaded communities. J Ecol 97:393–403
Irl S, Jentsch A, Walther GR (2013) *Papaver croceum* Ledeb.: a rare example of an alien species in alpine environments of the Upper Engadine, Switzerland. Alp Bot 123:21–30
Johnston FM, Pickering CM (2001) Alien plants in the Australian Alps. Mt Res Dev 21:284–291
Körner C (2002) Mountain biodiversity, its causes and function: an overview. In: Körner C, Spech EM (eds) Mountain biodiversity: a global assessment. Pathenon, London, pp 3–20
Körner C (2007) The use of "altitude" in ecological research. Trends Ecol Evol 22:569–574
Kueffer C, Alexander J, Daehler C, McDougall K, Pauchard A, MIREN Consortium http://www.miren.ethz.ch/people/ (2008) The Mountain Invasion Research Network (MIREN). Mt Res Initiat Newsl 1:12–14
Lowe S, Browne M, Boudjeles S, De Poorter M (2000) 100 of the world's worst invasive alien species. A selection from the Global Invesive Species Database. The invasive species specialist group of the species survival commission (SSC) of the World Conservation Union (IUCN)

Mallen-Cooper J, Pickering CM (2008) Linear declines in exotic and native plant species richness along an increasing altitudinal gradient in the Snowy Mountains, Australia. Austral Ecol 33:684–690

Manchester SJ, Bullock JM (2000) The impacts of non-native species on UK biodiversity and the effectiveness of control. J Appl Ecol 37:845–864

Marini L, Battisti A, Bona E, Federici G, Martini F, Pautasso M, Hulme PE (2011) Alien and native plant life-forms respond differently to human and climate pressures. Glob Ecol Biogeogr 21:534–544

Marini L, Gaston KJ, Prosser F, Hulme P (2009) Contrasting response of native and alien plant species richness to environmental energy and human impact along alpine elevation gradients. Glob Ecol Biogeogr 18:652–661

Marini L, Bertolli A, Bona E, Federici G, Martini F, Prosser F, Bommarco R (2013) Beta-diversity patterns elucidate mechanisms of alien plant invasion in mountains. Glob Ecol Biogeogr 22:450–460

McDougall KL, Morgan JW, Walsh NG, Williams RJ (2005) Plant invasions in treeless vegetation of the Australian Alps. Perspect Plant Ecol Syst 7:159–171

McDougall KL, Alexander JM, Haider S, Pauchard A, Walsh NG, Kueffer C (2011a) Alien flora of mountains: global comparisons for the development of local preventive measure against plant invasions. Divers Distrib 17:103–111

McDougall KL, Khuroo AA, Loope LL, Parks CG, Pauchard A, Reshi ZA, Rushworth I, Kueffer C (2011b) Plant invasions in mountains: global lessons for better management. Mt Res Dev 31 (4):380–387

Mincheva T, Barni E, Varese GC, Brusa G, Cerabolini G, Siniscalco C (2014) Litter quality, decomposition rates and saprotrophic mycoflora in *Fallopia japonica* (Houtt.) Ronse Decraene and in adjacent native grassland vegetation. Acta Oecol 54:29–35

Nogués-Bravo D, Araújo MB, Romdal T, Rahbek C (2008) Scale effects and human impact on the elevational species richness gradients. Nature 453:216–220

Parks CG, Radosevich SR, Endress BA, Naylor BJ, Anzinger D, Rew LJ et al (2005) Natural and land use history of the Northwest mountain ecoregions (USA) in relation to patterns of plant invasion. Perspect Plant Ecol 7:137–158

Pauchard A, Alaback PB (2004) Influence of elevation, land use, and landscape context on patterns of alien plant invasions along roadsides in protected areas of south-central chile. Conserv Biol 18:238–248

Pauchard A, Kueffer C, Dietz H, Daeler CC, Alexander J, Edwards P et al (2009) Ain't no mountain high enough: plant invasion reaching new elevations. Front Ecol Environ 7:479–486

Pearson RG, Dawson TP (2003) Predicting the impacts of climate change on the distribution of species: are bioclimate envelope models useful? Glob Ecol Biogeogr 12:361–371

Pickering CM, Hill W (2007) Roadside weeds of the Snowy Mountains, Australia. Mt Res Dev 27:359–367

Pyšek P, Jarosik V, Kucera T (2002) Patterns of invasion in temperate nature reserves. Biol Conserv 104:13–24

Rejmánek M (1989) Invasibility of plant communities. In: Drake JA, Mooney HA, di Castri F, Groves RH, Kruger M, Rejmánek M, Williamson M (eds) Biological invasions. A global perspective. Wiley, Chichester, pp 369–388

Rejmánek M, Randall JM (1994) Invasive alien plants in California: summary and comparison with other areas in North America. Madrono 41:161–177

Richardson DM, Thuiller W (2007) Home away from home—objective mapping of high-risk source areas for plant introductions. Divers Distrib 13:299–312

Siniscalco C, Barni E, Bacaro G (2011) Alien species distribution along the elevation gradient in the western Italian Alps. Plant Biosyst 145:150–158

Stohlgren TJ, Binkley D, Chong GW, Kalkhan MA, Schell LD, Bull KA et al (1999) Exotic plant species invade hot spots of native plant diversity. Ecol Monogr 69:25–46

Tassin J, Riviere NJ (2003) Species richness altitudinal gradient of invasive plants on Reunion Island (Mascareigne Archipelago, Indian Ocean). Rev Ecol-Terre Vie 58:257–270

Tecco PA, Diaz S, Cabido M, Urcelay C (2010) Functional traits of alien plants across contrasting climatic and land-use regimes: do aliens join the locals or try harder than them? J Ecol 98:17–27

Theoharides KA, Dukes JS (2007) Plant invasion across space and time: factors affecting nonindigenous species success during four stages of invasion. New Phytol 176(2):256–273

Thuiller W, Richardson DM, Pyšek P, Midgley GF, Hughes GO, Rouget M (2005) Niche-based modelling as a tool for predicting the risk of alien plant invasions at a global scale. Glob Chang Biol 11:2234–2250

Vacchiano G, Barni E, Lonati M, Masante D, Curtaz A, Tutino S, Siniscalco C (2013) Monitoring and modeling the invasion of the fast spreading alien *Senecio inaequidens* DC. in an alpine region. Plant Biosyst 147:1139–1147

Vaclavik T, Meentemeyer RK (2009) Invasive species distribution modeling (iSDM): are absence data and dispersal constraints needed to predict actual distributions? Ecol Model 220:3248–3258

Walter J, Essl F, Englisch T, Kiehn M (2005) Aliens in Austria: habitat preferences and ecological effects. Neobiota 6:13–25

Walther GR, Roques A, Hulme PE, Sykes MT, Pyšek P, Kühn I et al (2009) Alien species in a warmer world—risks and opportunities. Trends Ecol Evol 24:686–693

Weiher E, Keddy PA (1995) Assembly rules, null models, and trait dispersion: new questions from old patterns. Oikos 74:159–164

Williamson M (2006) Explaining and predicting the success of invading species at different stages of invasion. Biol Invasions 8:1561–1568

Wilson JRU, Richardson DM, Rouget M, Proche S, Amis MA, Henderson L, Thuiller W (2007) Residence time and potential range: crucial considerations in modelling plant invasions. Divers Distrib 13:11–22

Timberline and Alpine Vegetation in the Northern Apennines: Bioclimate Scenery and Vegetation Diversity

6

Carlo Ferrari

Abstract

From Mt. Orsaro to Mt. Corno alle Scale, the Northern Apennines include some of the highest peaks of the whole chain. The areas above the timberline are chiefly above 1700 m asl, and corresponds to the altitudinal limit of beech forests, which appears as an abrupt timberline. They are rare and scattered on the highest peaks areas above the timberline are can be considered an "archipelago" of alpine islands.

As suggested by the known beech thermal requirements, the present upper timberline (1825 m asl) corresponds to a mean annual temperature of 4.5 °C and to the warmest month (July) mean temperature of 13 °C. The July mean temperature corresponds to that of maritime mountains such as the Appalachians and the New Zealand Alps where a deciduous timberline can be found. Thermal data do not take into account site factors, such as land morphology, wind exposure and snow cover duration which create a high habitat and vegetation diversity. Our available knowledge on vegetation diversiy is based on descriptions following the Braun-Blanquet approach. Vegetation is characterized by ericaeous dwarf shrublands (*Rododendro-Vaccinietalia*), grasslands (*Caricetalia curvulae*; *Nardetalia*), marshes with reeds (*Phragmitetalia*) and sedges (*Caricetalia nigrae*), rock faces (*Potentilletalia caulescentis*), talus slopes (*Thlaspietalia stylosi*, *Androsacetalia alpinae*), springs and streams (*Cratoneuro-Philonotidetalia*), snow beds (*Salicetalia herbaceae*).

Keywords

Alpine ecology • Biodiversity • Biogeography • Phytosociology • Vegetation ecology

C. Ferrari (✉)
Department of Biological, Geological and Environmental Sciences, Bologna University, via Irnerio 42, 40126 Bologna, Italy
e-mail: carlo.ferrari@unibo.it

F. Pedrotti (ed.), *Climate Gradients and Biodiversity in Mountains of Italy*, Geobotany Studies, https://doi.org/10.1007/978-3-319-67967-9_6

6.1 Introduction

The crucial role of the transition areas between phytogeographic regions with respect to the genetic and taxonomic plant diversity has long been known. This is specially true for the areas above the timberline on the southernmost limit of Central European region where the alpine belt is rare and scattered on the highest peaks. This is the case in the few summit areas that can be found on the Northern Apennines (Pignatti 1994), a chain which geographically belongs to the south European mountains. The close proximity of their SW slopes to the Mediterranean Sea leads to marked climatic differences between the SW and NE slopes of the chain (Zanella 1989).

During the late Pleistocene the high north-eastern slopes of the Northern Apennines were marked by small glaciers which have left very clear traces (glacial rings and moraine deposits) and by periglacial processes, which are still active (Losacco 1982).

The natural history of the NE exposed summits and their close proximity to the Maritime Alps have given rise to an alpine flora very similar to that of the southwestern end of the Alps, although significantly impoverished (Ferrarini 1982; Foggi 1990; Pignatti 1994; Tomaselli and Agostini 1994). Boreal and arctic-alpine taxa are distributed in different communities usually covering small areas. Some plant communities (e.g. snow beds and *Caricion curvulae* grasslands) reach on the chain one of the southwestern end of their geographic distribution in Europe. Such fact underlines the importance of the Northern Apennines as a deposit of genetic and taxonomic plant diversity as well as of some aspects of the biological history of the south European mountains.

6.2 The Study Area

The Northern Apennines are a mountain system nearly 150 km long, separating the Po plain northwards from the Italian peninsula southwards, approximately between 44°–44°30′ N and 9°45′–11° E. The highest mountains lie between the Cisa pass northwards and the valley of Reno river southwards. In this sector only few summits exceed 2000 m asl, the highest peak being Mt. Cimone (2165 m) and Mt. Cusna (2120 m). The timberline corresponds to the altitudinal limit of beech woods, between about 1700 and 1900 m asl. As usual when the upper forest belt is dominated by broad-leaved deciduous tree (e.g. Crawford 1989; Wardle 1986; Cogbill and White 1991; Holtmeier 2013) there is an abrupt timberline. The areas above the timberline can be considered an "archipelago" of alpine islands (Fig. 6.1). The arenaceous bedrocks consist of a turbiditic compact textured sandstone (upper Oligocene—lower Miocene) locally known as "Macigno" formation (Dallan Nardi and Nardi 1974). The northern slopes between the Cisa Pass and the Mt. Corno alle Scale, are characterized by glacial landforms where periglacial processes are still active (Losacco 1982; Carton and Panizza 1988).

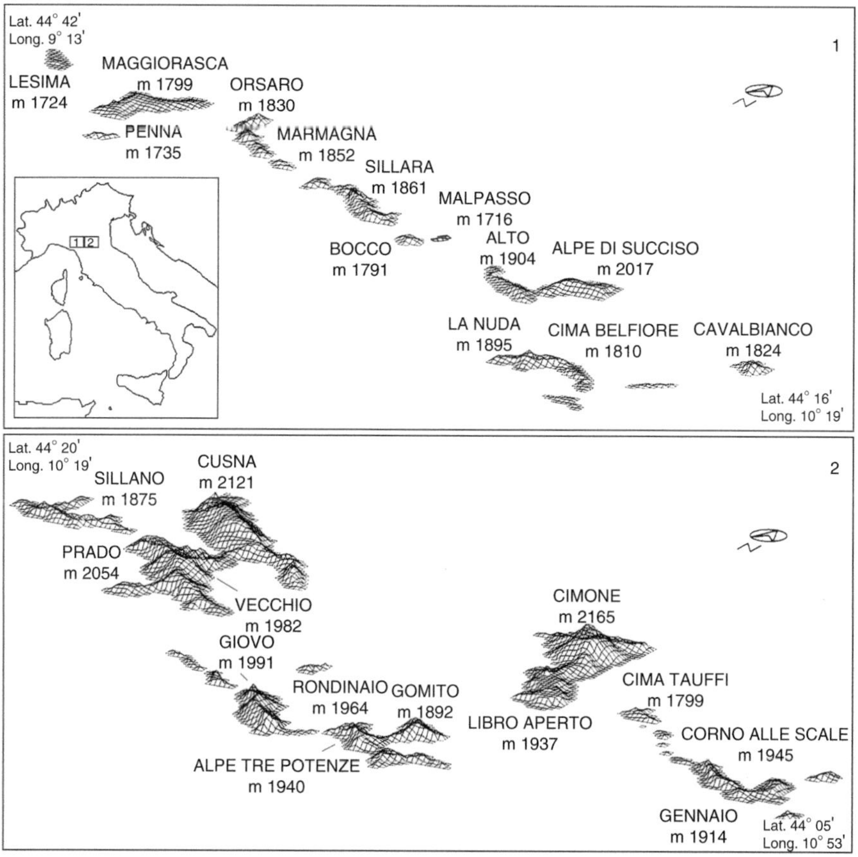

Fig. 6.1 The Northern Apennine "alpine islands" above the timberline (Ferrari and Piccoli 1997)

6.3 Timberline Distribution and Its Thermal Requirements

As shown by Pezzi et al. (2008) the Northern Apennine natural timberline lies between about 1700 m asl and 1825 m asl. Its most frequent adjacencies are with *Vaccinium* dwarf shrublands (*Rhododendro-Vaccinietalia*). The upper timberline (1825 m asl) corresponds to some thermal requirements, such as (1) a mean annual temperature of 4.5 °C; (2) a coldest month mean temperature of −2.3 °C; (3) a summer mean (June–September) of 11.5 °C; (4) warmest month (July) mean temperature of 13 °C; (5) 139 days with maximum temperatures of 10 °C or higher.

The July mean temperature corresponds to that of maritime mountains such as the Appalachians and the New Zealand Alps (Grace 1977; Crawford 1989) where a deciduous timberline can be found (Fig. 6.2). The geographical location of the Northern Apennines close to the Mediterranean Sea seems to confirm this maritime or "suboceanic" bioclimate characterization.

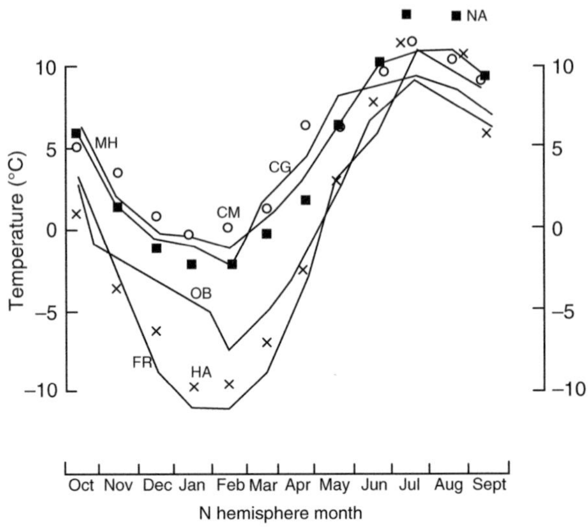

Fig. 6.2 The calculated annual trend (from October to September) of mean air temperature at the highest Northern Apennine timberline (NA, *filled square*, 1825 m). The Apennine trend (Pezzi et al. 2008) is compared with the same trend at different geographical locations (Grace 1977). The locations are identified by initials as follows: CG—Cairngorms, Scotland (760 m); CM—Craigieburn Mountains, New Zealand, (*open circle*, 1340 m); FR—Front Range Colorado, USA (3350 m); HA—Haugastol, Handargervida, Norway (multiplication sign, 988 m); MH—Moor House, northern England (558 m); OB—Obergurgle, Austria (2070 m)

These thermal features do not take into account the some site factors, such as wind-exposed topography and snow cover duration (Tomaselli 1991; Ferrari and Rossi 1995): which play a crucial role in the Northern Apennine habitat diversity.

6.4 Vegetation Diversity

The summit Northern Apennine landscape is characterized by various plant communities selected by landforms, snow cover duration and soil features. With reference to vegetation diversity the following habitats can be describe:

- heaths
- grasslands
- snowbeds
- wetlands
- rock faces
- talus slopes

Vegetation and soil features of some selected plant communities was described by Dowgiallo et al. (1998). On Mt. Prado (2054 m asl) Ferrari and Rossi (1995) described

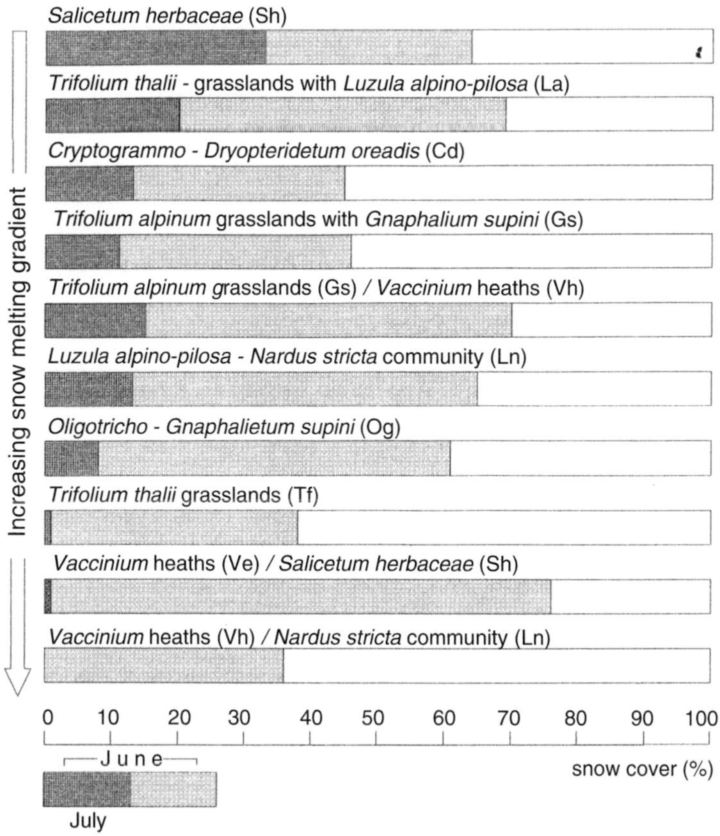

Fig. 6.3 The ten plant communities characterized by the longest snow cover duration in the Northern Apennines. The June and July snow cover (%) are shown (Ferrari and Rossi 1995)

the relationship between some rare plant communities and the snow cover duration (Fig. 6.3). The soil features was described in detail by Tomaselli (1991).

6.4.1 Heaths

Ericaceous dwarf shrublands dominated by *Vaccinium uliginosum* ssp. *microphyllum* (= *V. gaultheriodes*) are widely distributed above the timberline and are considered the climax vegetation of the Northern Apennine summit vegetation (Pignatti 1994). The early descriptions of this vegetation dates back to Lüdi (1943), Sarfatti and Pedrotti (1966), Ferrari (1978) and Barbero and Bonin (1980), which compared this vegetation to the *Empetro -Vaccinietum* from the Alps, underlying its species richness decrease, as usual for the plant communities at the margin of their geographic distribution (Braun-Blanquet 1964).

Ferrari and Piccoli (1997) described three main vegetation types corresponding to different microclimate features: a) *Empetro-Vaccinietum gaultheriodis* subass. *juncetosum trifidi*, which corresponds to sites with a duration of snow cover less than 6 months (November–April) and where the maximum speed of the prevailing winter wind (north-easternly) is around 17 m/s; b) *Rhododendretum ferruginei*, which is a very rare, sometimes fragmentary, community (e.g. Ferrarini 1973, 1974) settled on small areas with a duration of a high snow cover of about 7 months (November–May) and c) *Hyperico richeri-Vaccinietum gaultherioidis*, which corresponds to sites with "suboceanic" climate, where the maximum winter wind speed is about 11 m/s (e.g. Tranquillini 1969; Ferrari and Rossi 1995). Figure 6.4 shows a tentative ecological diagram of the three plant communities.

Both types of *Vaccinium* heaths are settled on colluvial deposits or morainic materials. Most soils are rather deep with more or less pronounced podzolic features (Dowgiallo et al. 1998).

These communities are dominated by boreal and south-European orophytes. A relevant set of arctic-alpine taxa is into the *Empetro -Vaccinietum gaultherioidis* subass. *juncetosum trifidi*. *Rhododendron ferrugineum*, which is widely distributed on the Alps, on the northern Apennines reach its southernmost distribution in Italy with rare small populations genetically similar to the populations from the Maritime Alps (e.g. Bruni et al. 2012). All the plant communities belong to the phytosociological order *Rhododendro-Vaccinietalia* and stress the phytogeographic connection among the Northern Apennine summit vegetation and the subalpine vegetation belt of the Alps. In particular, the *Empetro-Vaccinietum gaultherioidis* subass. *juncetosum trifidi* contains the most arctic alpine taxa and it is to be included in the alliance *Loiseleurio-Vaccinion*. The other two plant communities belong to the alliance *Rhododendro-Vaccinion*.

6.4.2 Grasslands

The most grasslands can be ascribed to the acidophytic association *Sileno excapae-Trifolietum alpini*, belonging to the alliance *Caricion curvulae* Braun-Blanquet 1925, with some subassociations and variants (Tomaselli and Rossi 1994), which correspond to different wind exposed slopes and duration of snow cover. The association is restricted to the Northern Apennines and is the southernmost outpost of the *Caricion curvulae* in Italy. Two species appear the most predictive for characterizing this vegetation: *Trifolium alpinum* and *Silene acaulis* ssp. *bryoides*. The *Caricion curvulae* species are *Senecio incanus* and *Hieracium glanduliferum*

Within the grasslands dominated by *Nardus stricta* three different vegetation types were described (Tomaselli 1994). The association *Geo montani-Nardetum* corresponds to heavily pastured acidophytic grasslands. It is characterized by a lot of diagnostic species of *Nardion* (e.g. *Geum montanum*, *Potentilla aurea*, *Gentiana acaulis*, *Polygala alpestris*) *Nardetalia* and *Nardo-Callunetea* (e.g. *Festuca nigrescens*, *Luzula multiflora*, *Potentilla erecta*). This vegetation occurs in low subalpine vegetation belt as well as in the high montane belt, where it corresponds to pastured grasslands derived from clearance of beech woods The association

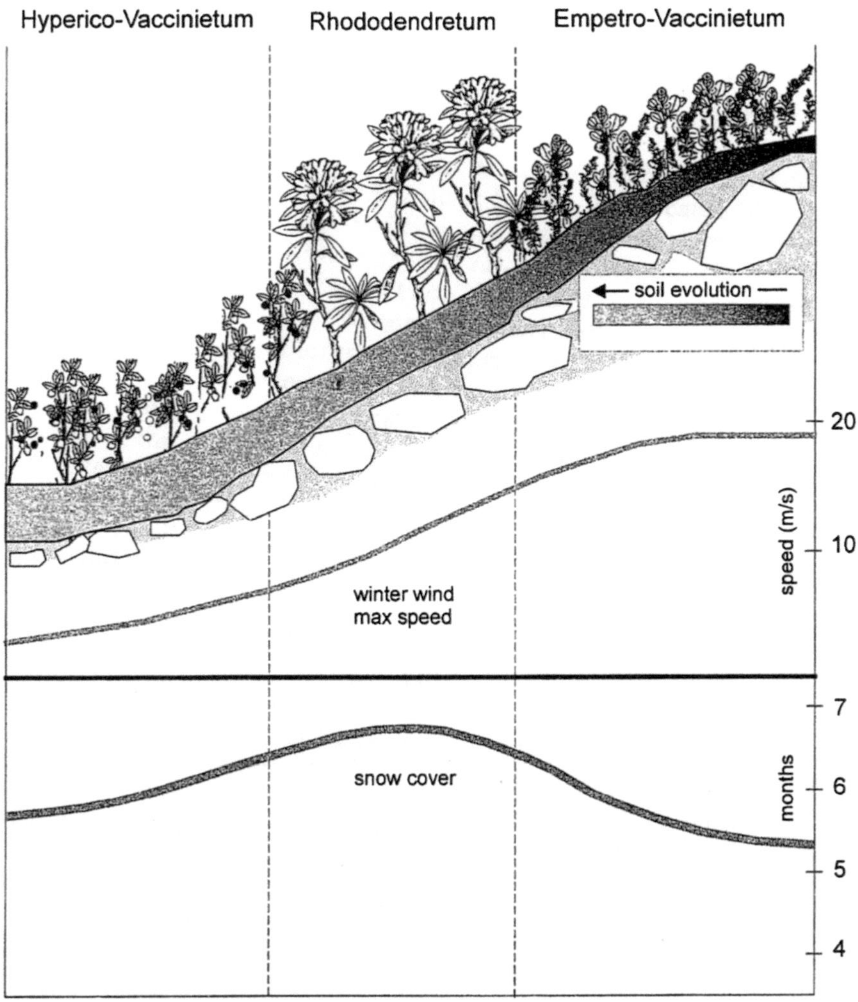

Fig. 6.4 A tentative ecological diagram of the three dwarf shrublands above the Northern Apennine timberline (Ferrari and Piccoli 1997)

Violo cavillieri-Nardetum (Credaro and Pirola 1975) corresponds to natural grasslands with a high content of *Juncetea trifidi* (e.g. *Trifolium alpinum*, *Phyteuma emisphaericum*, *Juncus trifidus*). A natural *Nardus* community with *Luzula alpino-pilosa* is settled on small hollows sited just below or along the summit ridges with a long duration of snow cover. This habitat is ecologically contiguous to snowbeds and the main feature of this community is the high occurrence of some *Salicetea herbaceae* species (e.g. *Luzula alpino-pilosa*, *Gnaphalium supinum*).

The ledges of the arenaceous rock walls are the habitat of the *Aquilegio-Anemonetum narcissiflorae* (Tomaselli 1994), the only mesophytic grassland as yet described in the high Northern Apennines

Grasslands dominated by *Brachypodium genuense* and *Festuca paniculata,* are frequent on southern exposed stony and scree slopes. These natural grasslands are characterized by *Nardo-Callunetea* species and lies on steep dry slopes with winter low soil temperatures, due to the short duration of snow cover. These grasslands are not to be confused with the high pastured grasslands, derived by the destruction of *Vaccinum* heaths and now dominated by *Brachypodium genuense*.

6.4.3 Snowbeds

In the Northern Apennines, which are a phytogeographic tension zone between the Eurosiberian and Mediterranean region, the correspondence between the chionophilous *Salicetea herbaceae* species and snowbeds remains. Nevertheless, it is somewhat weakened, because the chionophilous species are significantly reduced with respect to the snowbeds in the Alps (Turner 1961; Tomaselli 1991; Petraglia and Tomaselli 2007).

Six plant associations belonging to the class *Salicetea herbaceae* were described: *Salicetum herbaceae*, *Oligotricho hercynici-Gnaphalietum supini*, *Luzuletum spadiceae*, *Polytrichetum sexangularis*, *Poo-Cerastietum cerastoidis*, *Caricetum foetidae*.

The *Salicetum herbaceae* is named from *Salix herbacea*, (i.e. the linnean "arbor minima") a chionophilous willow which is distributed throughout the Alps (Aeschimann et al. 2004) but rarely occurs in the Northern and Central Apennines (Ferrarini 1974; Blasi et al. 2003). In the Northern Apennines Petraglia and Tomaselli (2007) found this association in eight stand. In addition to *Salix herbacea*, characteristic species are *Leucanthemopsis alpina*, *Silene suecica*, and *Senecio incanus*. *Poa alpina*, *Agrostis rupestris* and *Sedum alpestre* occur most frequently as well as some *Caricetea curvulae* species, but usually with low cover. The moss layer contains *Polytrichum juniperinum* and *P. piliferum*. A typical *Salicetum herbaceae* is restricted to sheltered sites on the summit slopes of Mt. Prado, where it occurs on well drained gravelly soils with some solifluctions. A *Sanionia uncinata* variant was described on the northern slope of Mt. Cimone.

The most common snowbed association in the Northern Apennines is the *Oligotricho hercynici-Gnaphalietum supini* which is entirely confined in the northern Apennines. It corresponds to very low and from open to close turfs in which mat-forming rosette herbs, trussocky grasses and moss carpets provide the most distinctive features. *Gnaphalium supinum* is the most abundant species with *Agrostis rupestris*, *Plantago alpina* and *Sedum rupestre*. The most common bryophytes are *Kiaeria starkei*, *Marsupella brevissima*, *Pohlia drummondii*.

The communities dominated by *Luzula alpino-pilosa* (*Luzuletum spadiceae*) can be found on sheltered sites at altitudes from 1770 and 2120, mainly within narrow and steep N-facing gullies at sheltered sites which retain snow cover long and where snow melt is the major source of water. The number, frequency and cover of *Salicetea herbaceae* species is strongly reduced from those from the Alps (Giacomini and Pignatti 1955; Giacomini et al. 1962) indicating that this association occurs in the Northern Apennines only as an impoverished form. *Luzula alpino-pilosa* is a chionophilous species showing high occurrence also in other snowbed associations but playing its most important role within this community. The other constant species is *Festuca violacea* ssp. *puccinellii*. The most frequent species are *Cryptogramma crispa*, *Hypericum richeri*, *Saxifraga rotundifolia*, *Athyrium distentifolium* and *Viola biflora*.

At only one sheltered site, within the fossil western glacial cirque of Mt. Giovo, Petraglia and Tomaselli (2007) found the association *Polytrichetum sexangularis*, a bryophyte dominated snowbed plant community. The most outstanding species in the moss carpet are *Polytrichastrum sexangulare* and *Marsupella brevissima*. The prominent vascular plants are *Nardus stricta* and *Festuca rubra* ssp. *commutata*.

On the northern slopes of Mt. Cusna and Mt. La Piella, at altitude between 1900 and 1970 m, on the flat bottoms of sheltered ill-drained hollows as well as the shores of a small lake of glacial origin, it can be found few stands of the association *Poo-Cerastietum cerastoidis*. *Cerastium cerastoidis* and *Pohlia ludwigii* are its character species, with *Poa supina*, *Plantago alpina* and *Polytrichum juniperinum* as constant species.

The association *Caricetum foetidae* is dominated by the tufted shoots of *Carex foetida*. The most frequent vascular species are *Nardus stricta*, *Deschampsia caespitosa*, *Crepis aurea* ssp. *glabrescens* and *Plantago alpina*. This association is restricted to the highest summits of the central part of the Northern Apennines where it colonizes sheltered flattened sites, lying at the bottom of small glacial cirques, at nivation hollows and at small hollows delimited by protalus ramparts. Soils are long waterlogged during the summer.

As a consequence of the shortening of the snow cover duration, the Northern Apennine snowbed communities are likely exposed to the risk of local extinction or of loss of some species with narrow ecological needs.

6.4.4 Wetlands

The vegetation of the Northern Apennine wetlands was described in detail by Gerdol and Tomaselli (1993).

The aquatic vegetation colonizing the central area of the small lakes mainly consists of the association *Sagittario-Sparganietum emersi*, dominated by *Sparganium emersum* populations. The inner lake edges are the habitat of the *Eleocharis palustris* and *Alopecurus aequalis* community, which have *Ranunculus tricophyllus* as a frequent species. The surrounding fens, where water pH is slightly lower and electrical conductivity higher than in the *Sagittario-Sparganietum emersi* are the habitat of a *Carex rostrata* and *Carex vesicaria* community, which is rich in *Scheuchzerio-Caricetea* species. *Menyanthes trifoliata* populations are frequent in this habitat.

The outer lake edge is the habitat of a community dominated by *Carex nigra* and *Eriophorum angustifolium (Caricetum nigrae)*.

The summit mire vegetation essentially consists of fens and *Sphagnum* mats ("Schwingrasen") and belongs to the order *Caricetalia nigrae*. Near and above the Northern Apennine timberline this vegetation includes plant communities of high phytogeographic and ecological interest.

The rare *Eriophoretum scheuchzeri*, can be found only in few sites (e.g. Mt. Sillara, Baccio Lake). This fact dramatically contrasts with what happens in the Alps, where this association is a widespread fen association in the alpine and subalpine belt (e.g. Credaro and Pirola 1975; Krisai and Peer 1980; Ellenberg 1987; Körner 2003). The *Drepanocladus exannulatus-Juncus filiformis* community is frequent at higher altitudes and colonizes pools originated by sandstone erosion. These pools are characterized by a compact peaty round which usually dries out in the summer and is frequently disturbed by grazing sheeps. Peat water is acidic and poor in nutrients. Snow may persist long, mainly in the sites sheltering snow-beds plants.

The relatively widespread *Caricetum nigrae*, corresponds to low sedge fens developed on a carpet of *Hypnaceae* and *Sphagnum* with some variants corresponding to different waterlogging, peat water pH and nutrient poorness. For its phytogeographic interest the most relevant vegetation type is the association *Sphagno (nemorei)–Caricetum nigrae*, including *Sphagnum compactum*, *Tricophorum caespitosum*, and the locally rare *Ptilidium ciliare*, *Polytrichum strictum* and *Sphagnum subnitens*. This association can be considered the main refuge for *Oxycocco-Sphagnetea* species in the Northern Apennines, as in other south European mountains, (Soò 1957; Vanden Berghen and Peeters 1982) where the climate hinders the development of raised *Sphagnum* hummocks and the *Oxycoccco-Sphagnetea* species become rare, mainly due to phytogeographic reasons.

The edge of spring and high stream waters are the habitat of the association *Chaerophyllo-Cardaminetum asarifoliae* (Gerdol and Tomaselli 1988), characterized by *Chaerophyllum hirstum* and *Cardamine asarifolia, Caltha palustris* ssp. *laevis*, *Carex frigida*, *Cratoneurum commutatum*, *Saxifraga stellaris*, *Bryum schleicheri* and *Saxifraga aizoides* are the most frequent species. The edge of stream waters with low organic content and pH close to neutrality is also the core habitat of the endemic *Cirsium spinosissimum* ssp. *bertolonii* whose present wide distribution on disturbed grasslands with high organic content is due to livestock settings.

6.4.5 Rock Faces and Talus Slopes

On arenaceous rock crevices three plant communities were described by Tomaselli (1994).

The association *Silenetum saxifragae*, is the most widespread and can be found only in the few arenaceaous rocks of the Apuan Alps. It is characterized by *Silene saxifraga*, with *Arenaria bertolonii. Globularia incanescens*, *Festuca robustifolia* as frequent species. It belongs to the alliance *Saxifragion lingulatae* dell'ordine *Potentilletalia caulescentis*.

The association *Asplenio—Cystopteridetum fragilis* which is characterized by the ferns *Cystopteris fragilis* and *Asplenium viride*, is settled on the sheltered and moist lower parts of N-exposed rocks. It belongs to the alliance *Cystopteridion* within the order *Potentilletalia caulescentis*.

The association *Draba aizoidis -Primuletum apenninae* is limited to the Northern Apennines and it is characterized by the endemic *Primula apennina*, the only Apennine primrose with red flowers, and *Draba aizoides*, which is locally settled only in the arenaceous rock crevices. The endemic *Primula apennina* is belonging to the sect. *Auricula* and along with other related species (*P. daonensis*, *P. cottia*, *P. pedemontana* and *P. villosa*) is probably derived from *P. irsuta*, as the most alpine species in the section (Crema et al. 2009). From a phytogeographic view point this association can be considered a southern vicariant of the alpine *Asplenio-Primuletum irsutae* which belongs to the alliance *Androsacion vandellii* within the order *Androsacetalia vandellii*.

Marly and marly areanceous talus slopes are colonized by the association *Arenarietum bertolonii*, characterized by *Arenaria bertolonii* and *Trifolium thalii*. This association is limited to the Tuscan-Emilian sector of the Northern Apennines. Owing to the calcareous filling of the arenaceous rocks, it can be considered the northernmost outpost of the calcareous scree vegetation from the Central Apennines.

The boulder slopes and block fields are colonized by the *Cryptogrammo-Dryopteridetum oreadis*, characterized by the ferns *Cryptogramma crispa* and *Dryopteris oreades*. *Polystichum lonchitis*, *Dryopteris filix-mas* and *Phegopteris connectilis* are locally frequent.

A *Pinus mugo ssp. uncinata* population can be found on the Mt. Nero (1799 m) and Mt. Maggiorasca (1804 m) ophiolitic stabilized screes. It is the only Northern Apennine population of this conifer, which is probably a relict of late Würm widespread populations.

6.5 Conclusion

The northern Apennines, geographically interposed between the Alps and the Central Apennines, are a phytogeographic tension zone between the Eurosiberian and Mediterranean regions (Foggi 1990; Tomaselli and Agostini 1994). On this southern European mountain chain the alpine belt is rare and scattered on the highest peaks. The few areas above the timberline are of high biogeographic and environmental interest, as they are "alpine islands" (Ferrari et al. 2000) where some boreal and arctic-alpine taxa find the southernmost border of their geographic distribution in Europe (e.g. Quezel 1953; Credaro and Pirola 1975; Gamisans 1977). The plant communities have different degree of phytogeographic connection with the siliceous Alps and their flora is generally impoverished with respect to comparable habitats of both northern and central Europe. Wetlands, snowbeds and heaths are the most striking examples on this regard. Nevertheless their importance as refuge habitats makes these plant communities sensitive areas with respect to climatic changes.

During the last 50 years a trend of reducing winter snowfall associated with a mean annual temperature increase was documented in the Northern Apennine summit areas. (Tomozeiu et al. 2002, 2003) As a consequence, a progressive shortening in the duration of snow cover could be expected in these areas. It could mostly affect the plant communities and species occurring on moderate slopes, such as snowbeds and their chionophilous species, the *Oxycocco-Sphagnetea* species which found on some mires their refuge habitat. These habitats and their plant communities could shrink or even disappear at some sites. As global climate changes can't be modified at a local scale, the best strategy is only a large nature conservation, so the most biological histories can coexist (Worster 1985).

Appendix

Syntaxonomical Vegetation Scheme

Class Asplenietea trichomanis (Br.-Bl. In Meier et Br.-Bl. 1934) Oberd.1977
Order Potentilletalia caulescentis Br. Bl. In Br.Bl. et Jenny 1926
Alliance Cystopteridion (Nordhag. 1936) J.L. Rich. 1972.
  Association Asplenio-Cystopteridetum fragilis Oberd. (1936) 1949.
Order Androsacetalia vandelli Br.-Bl. In Meier et Br.-Bl. 1934
Alliance Androsacion vandellii Br.-BL. In Br.-Bl. Et Jenny 1926
  Association Drabo aizoidis-Primuletum apenninae Tomaselli 1994.

Class Thlaspietea rotundifolii Br.-Bl. Et al. 1947.
Order Thlaspietalia stylosi Avena et Bruno 1975
Alliance Linario -Festucion dimorphae Avena et Bruno 1975
  Associatiion Arenarietum bertolonii Credano et Pirola 1975
Order Androsacetalia alpinae Br. Bl. In Br.-Bl. et Jenny 1026
Alliance: Dryopteridion oreadis Rivas-Martinez 1977
  Association Cryptogrammo-Dryopteridetum oreadis Rivas-Martinez in Rivas Martinez et Costa 1970

Class Montio-Cardaminetea Br.-Bl. et Tx 1943
Order Cratoneuro-Philonotidetalia Geissl. 1976
Alliance Cratoneurion commutati W. Koch 1928
  Association Chaerophyllo-Cardaminetum asarifoliae Gerdol et Tomaselli 1988

Class Salicetea herbaceae Br. Bl. et al. 1947
Order Salicetalia herbaceae Br.—Bl. in Br.-Bl. et Jenny 1926
Alliance Salicion herbaceae Br.—Bl. in Br.-Bl. et Jenny19 26
Associations Salicetum herbaceae Rübel 1912; Oligotricho hercynici-Gnaphalietum supini Tomaselli 1991; Luzuletum spadiceae Rübel 1911, Polytrichetum sexangularis, Poo-Cerastietum cerastoidis Tomaselli 1991; Caricetum foetidae Frey 1922.

Class Phragmitetea Tx et Prs 1942
Order Phragmitetalia W. Koch 1926
Alliance Phragmition australis W. Koch. 1926
Association Sagittario-Sparganietum emersi Tx 1953
Alliance Magnocaricion W. Koch. 1926
Associations Carex rostrata community, Caricetum vesicariae Chouard 1924; Menyanthes trifoliata community

Claas Scheuchzerio-Caricetea nigrae (Nordh 1936) Tx 1937
Order Caricetalia nigrae (W. Koch 1926) Nordh. 1926
Alliance Caricion nigrae W. Koc 1926 em. Klika 1934.
Associations Eriophoretum scheuchzeri Fries 1913; Caricetum nigrae Br.-Bl. 1915; Sphagno nemorei - Caricetum nigrae Gerdol et Tomaselli 1993; Drepanocladus exannulatus and Juncus filiformis community.

Class Montio-Cardaminetea Br. -Bl. et R. Tx ex Klika et Hadac 1944 em. Zechmeister 1993
Order Montio-Cardaminetalia Paswloski 1928 em. Zechmeister 1993
Alliance Cratoneurion commutati Koch 1928
Association Chaerophyllo-Cardaminetum asarifoliae Gerdol et Tomaselli 1988.

Class Caricetea curvulae Br.-Bl. 1948
Order Caricetalia curvulae Br.-Bl. 1926 em Krajina 1933
Alliance Caricion curvulae Br.-Bl. 1925
Association Sileno excapae-Trifolietum alpini Tomaselli et Rossi 1994

Class Nardo-Callunetea Prsg 1949
Order Nardetalia Oberd. 1948 em Prsg 1949
Alliance Nardion Br.-Bl. in Br.-Bl. et Jenny 1926.

Associations Geo montani-Nardetum Ludi 1948: Violo cavillierii-Nardetum Credaro et Pirola 1975 corr.Tomaselli 1994; Brachypodium genuense community

Class Loiseleurio-Vaccinietea Eggler 1952
Order Rododendro-Vaccinietalia Br.-Bl. in Br.-Bl. et Jenny 1926
Alliance Loiseleurio -Vaccinion Br.-Bl. in Br.-Bl. et Jenny 1926
Association Empetro-Vaccinietum gaultherioidis Br.-Bl. in Br.-Bl. et Jenny 1926 corr. Grabherr 1993.
Alliance Rododendro-Vaccinion Br.-Bl. 1926
Association Hyperico richeri-Vaccinietum gaultherioidis Pirola et Corbetta 1971 corr. Ferrari 1997.

References

Aeschimann D, Lauber R, Moser DM, Theurillat JP (2004) Flora alpina. Zanichelli, Bologna
Barbero M, Bonin G (1980) La végétation de l'Apennin septentrional. Essai d'interpretation synthetique. Ecol Mediterr 5:273–313
Blasi C, Di Pietro R, Fortini P, Catonica C (2003) The main plant community types of the alpine belt of the Apennine chain. Plant Biosyst 137(1):87–110
Braun-Blanquet J (1964) Pflanzensoziologie, 3 Auflage edn. Springer, Wien
Bruni I, De Mattia F, Labra M, Grassi F, Fluch S, Berenyi M, Ferrari C (2012) Genetic variability of relict *Rhododendron ferrugineum* L. populations in the Northern Apennines with some inferences for a conservation strategy. Plant Biosyst 146(1):24–32
Carton A, Panizza M (1988) Il paesaggio fisico dell'alto Appennino emiliano. Grafis, Bologna
Cogbill CV, White PS (1991) The latitude-elevation relationship for spruce-fir forest and timberline along the Appalachian mountain chain. Vegetatio 94:153–175
Crawford RMM (1989) Polar and altitudinal limits to tree survival. In: Crawford RMM (ed) Studies in plant survival. Ecological case histories of plant adaptation to adversity. Blackwell, Oxford, pp 77–103
Credaro V, Pirola A (1975) Note sulla vegetazione ipsofila dell'Appennino tosco-emiliano. Atti Ist Bot Lab Critt Univ Pavia s.6, 10:35–58
Crema S, Cristofolini G, Rossi M, Conte L (2009) High genetic diversity detected in the endemic *Primula apennina* Widmer (Primulaceae) using ISSR fingerprinting. Plant Syst Evol 280:29–36
Dallan Nardi D, Nardi R (1974) Schema stratigrafico e strutturale dell'Appennino settentrionale. Mem Accad Lunig Sci 12:1–212
Dowgiallo G, Rossi G, Tomaselli M (1998) Vegetation and soil diversity in the areas above timberline of the Tuscan-Emilian Apennines (N Italy). Ecologie 29(1–2):159–162
Ellenberg H (1987) Vegetation ecology of Central Europe, 4th ed. Cambridge University Press, Cambridge
Ferrari C (1978) Elementi per una tipologia fitosociologica dei vaccinieti extrasilvatici dell'Appennino settentrionale. Arch Bot Biogeogr Italy 54:3–15
Ferrari C, Piccoli F (1997) The ericaceous dwarf shrublands above the Northern Apennine timberline. Phytocoenologia 27:53–76
Ferrari C, Rossi G (1995) Relationships between plant communities and late snow melting on Mount Prado (Northern Apennines, Italy). Vegetatio 120:49–58

Ferrari C, Portanova A, Pezzi G (2000) Plant diversity in some alpine islands of the Northern Apennines (Italy). Ber Reinh-Tüxen-Ges 12:399–404

Ferrarini E (1973) *Rhododendron ferrugineum* L. in fitocenosi relitte nell'Appennino settentrionale. Giorn Bot Ital 107:143–156

Ferrarini E (1974) Altre cenosi relitte reperite nell'Appennino settentrionale (ad *Antennaria carphatica*, a *Rhododendron ferrugineum*, a *Salix herbacea*, a *Saussurea discolor*). Webbia 29:105–112

Ferrarini E (1982) Note fitogeografiche sull'Appennino settentrionale nei suoi rapporti con le Alpi marittime. Lav Soc Ital Biogeogr Ns 9:211–226

Foggi B (1990) Analisi fitogeografica del distretto appenninico tosco-emiliano. Webbia 44 (2):169–196

Gamisans J (1977) La végétation des montagnes corses. Deuxiéme partie. Phytocoenologia 4:35–131

Gerdol R, Tomaselli M (1988) Phytosociology and ecology of strem vegetation in the summit region of the Northern Apennines. Boll Mus Storia Nat Lunigiana, Aulla 6–7:88–93

Gerdol R, Tomaselli M (1993) The vegetation of wetlands in the northern Apennines. Phytocoenologia 21(4):421–469

Giacomini V, Pignatti S (1955) Flora e vegetazione dell'alta Valle del Braulio, con particolare riferimento ai pascoli d'altitudine. Mem Soc Ital Sci Nat 11:47–238

Giacomini V, Pirola A, Wikus E (1962) I pascoli dell'alta Valle di S. Giacomo (Spluga). Flora et Vegetatio italica, 4. Gianasso, Sondrio

Grace J (1977) Climatic tolerance and the distribution of plants. New Phytol 106(suppl):113–130

Holtmeier FK (2003) Mountain timberlines. Ecology, patchiness, and dynamics. Kluwer, Dordrecht

Körner C (2003) Alpine plant life. Functional plant ecology of high mountain ecosystems. Springer, Berlin

Krisai R, Peer T (1980) Vegetationskundlich-ökologische Untersuchungen an drei Ostalpenmooren. Verh Zool-Bot Ges Wien 118(119):38–73

Losacco U (1982) Gli antichi circhi glaciali dell'Appennino settentrionale. Studio morfologico e paleogeografico. Atti Soc Nat Mat Modena 113:1–224

Lüdi W (1943) Uber rasengesellschaften und alpine Zwergstraucheiden in den Gebirgen des Apennin. Ber Geobot Forsch Inst Rubel Zurich 42:23–68

Petraglia A, Tomaselli M (2007) Phytosociological study of the snowbed vegetation in the Northern Apennines (Northern Italy). Phytocoenologia 37(1):67–98

Pezzi G, Ferrari C, Corazza M (2008) The altitudinal limit of beech woods in the Northern Apennines (Italy). Its spatial pattern and some thermal inferences. Folia Geobot 43:447–459

Pignatti S (1994) The climax vegetation above timberline in the northern and central Apennines. Fitosociologia 26:5–17

Quezel P (1953) Contribution a l'étude phytosociologique et géobotanique de la Sierra Nevada. Mem Soc Brot (Coimbra) 9:5–78

Sarfatti G, Pedrotti F (1966) I vaccinieti a *Vaccinium uliginosum* dell'Appennino tosco-emiliano. Giorn Bot Ital 73:333–334

Soò R (1957) Pflanzengesellschaften aus Bulgarien 1. Ann Univ Sci Budapest 1:231–239

Tomaselli M (1991) The snow-bed vegetation in the Northern Apennines. Vegetatio 94:177–189

Tomaselli M (1994) The vegetation of summit rock faces, talus slopers and grasslands in the Northern Apennines. Fitosociologia 26:35–50

Tomaselli M, Agostini N (1994) A comparative phytogeographic analysis of the summit flora of the Tuscan-Emilian Apennines and of the Apuan Alps. Fitosociologia 26:99–109

Tomaselli M, Rossi G (1994) Phytosociology and ecology of *Caricion curvulae* vegetation in the northern Apennines. Fitosociologia 26:51–62

Tomozeiu R, Lazzeri M, Cacciamani C (2002) Precipitation fluctuations during winter season from 1960-1995 over Emilia-Romagna, Italy. Theor Appl Climatol 72:221–229

Tomozeiu R, Pavan V, Cacciamani C (2003) Cambiamenti climatici in valori medi ed estremi di temperatura e precipitazioni in Emilia-Romagna. Quaderno Tecnico Arpa 11. Regione Emilia Romagna, Bologna

Tranquillini W (1969) Photosynthese und Transpiration einiger Holzarten bei verschieden strakem Wind. Zentralbl Gesamte Forstwes 86:35–48

Turner H (1961) Ökologische Untersuchungen in der subalpinen Stufe. 8- Die Niederschlags- und Schneeverhältnisse. Mitt Forstl Bundesversuchsanst Mariabrunn 59:265–315

Vanden Berghen C, Peeters A (1982) La végétation des sols mouilleés ou tourbeux de l'étage subalpin à Andorre (Pyrenées orientales). Bull Soc R Bot Belg 115:181–197

Wardle P (1986) Flora and fauna of Australasia, ages and origins. CSIRO, Melbourne, pp 45–61

Worster D (1985) Nature's economy. A history of ecological ideas. Cambridge University Press, Cambridge

Zanella G (1989) Il clima montano in Emilia-Romagna. In: Goidanich G, Gambi L, Ferrari C, Ricci Lucchi F (eds) Il mondo della natura in Emilia-Romagna. La montagna. Silvana editoriale, Milano, pp 141–164

7 Diachronic Analysis of Beech Forest in the Nebrodi Park (Northern Sicily)

G. Bazan, G. Castellano, S. Ciccarello, P. Marino, and R. Schicchi

Abstract

A comparison between the current surface of Beech forest in the Nebrodi Park (Northern Sicily) with the measurements performed in the past is here reported. 12.854 hectares of Beech forest were recognized against 10.336 hectares of 1959 with a percentage increase of 24.37%. The increase of forest is mainly due at the change of land use through a better forest management and grazing.

Keywords

Beech forest • *Fagus* • GIS • Sicily

7.1 Introduction

The European beech wood (*Fagus sylvatica* L.) covers most of the Old World's land mass, covering a broad biometeorological zone from temperate to warm climates (Ellemberg 1996).

Representatives of the family *Fagaceae* played an important role in the composition of the European Tertiary flora. As a principal component of biodiversity throughout the middle-late Tertiary, they are considered of great significance for the reconstruction of palaeoecosystems and for understanding their evolution in time and space, as well as the internal ecosystem dynamics. All this is due to the fact that many *Fagaceae* predominated in Tertiary forest (Palamarev and Mai 1998).

G. Bazan • R. Schicchi (✉)
Dipartimento di Scienze Agrarie e Forestali, Università di Palermo, Palermo, Italy
e-mail: rosario.schicchi@unipa.it

G. Castellano • S. Ciccarello • P. Marino
Dipartimento STEBICEF/Sezione di Botanica ed Ecologia vegetale, Università di Palermo, Palermo, Italy

F. Pedrotti (ed.), *Climate Gradients and Biodiversity in Mountains of Italy*, Geobotany Studies, https://doi.org/10.1007/978-3-319-67967-9_7

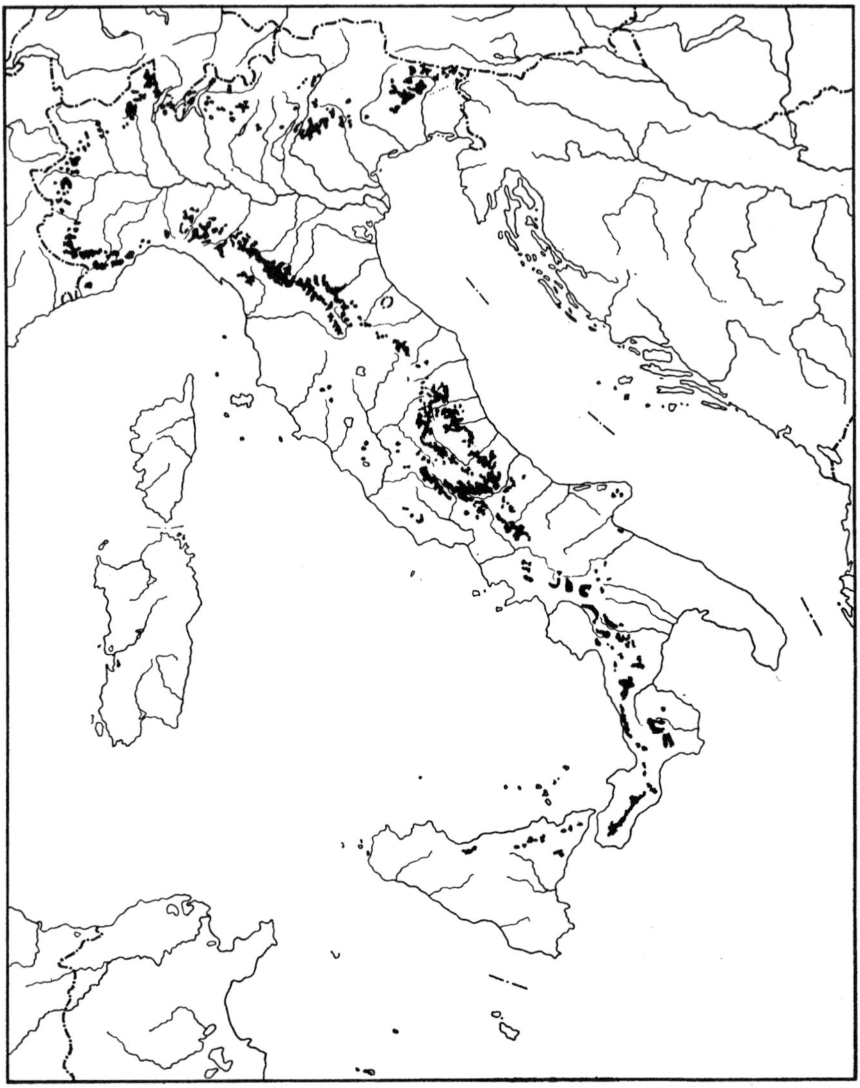

Fig. 7.1 Distribution of *Fagus sylvatica* in Italy (by Fenaoli 1970)

In Italy, according Nocentini (2009), Beech forests are present in all the regions except for Sardinia. In the Alps, beech generally forms pure stands above 1000 m altitude in areas with relatively low rainfall, while it grows at 600–700 m in more humid areas. On the Apennine mountains beech usually grows above 900–1000 m (Fig. 7.1). Beech forests are more widespread on the northern slopes and where rain and fog maintain moist air conditions. On the sunnier and warmer southern slopes, the lower vegetation limit for beech tends to move higher (Hofmann 1961; Pignatti

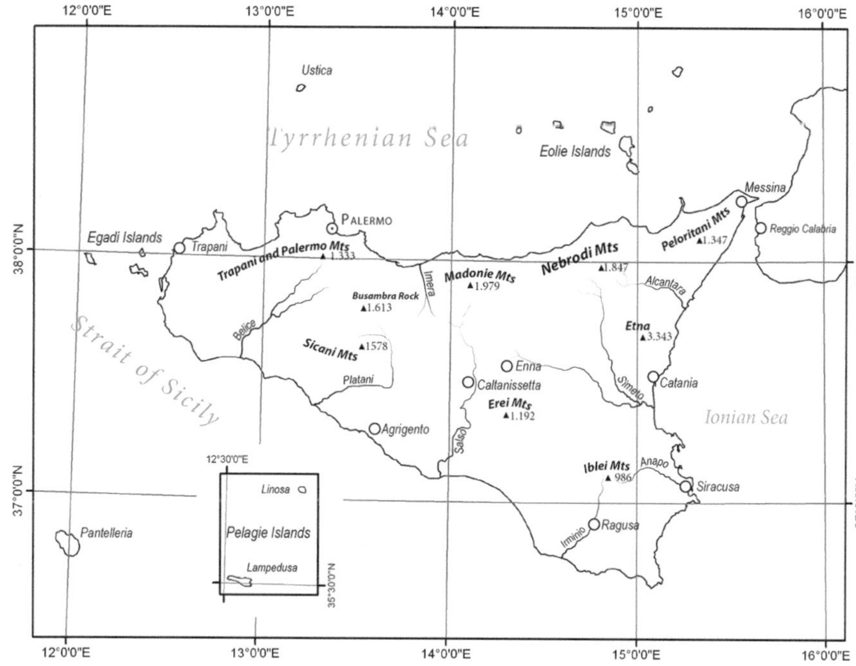

Fig. 7.2 Map of Sicily with the main localities

1998). On the northern slope of the Mount Etna in Sicily beech reaches an altitude of 2000 m (Hofmann 1960; Del Favero 2008).

In Sicily Beech forests are a relevant example of Cenozoic flora widespread in the Madonie Mts, Nebrodi Mts and Etna Volcano (Fig. 7.2) (Bazan et al. 2010). On the Etna Volcano, in the Lower Humid horizon, of the Supramediterranean belt, the Beech vegetation is represented by *Epipactido meridionalis-Fagetum sylvaticae*, floristically very poor and with a scarce shrub layer. On carbonatic substrata of Madonie Mts, beech forests are represented by *Luzulo siculae-Fagetum sylvaticae*, while on siliceous substrata of Madonie and Nebrodi, the mature stage of vegetation series is *Anemono apenninae-Fagetum sylvaticae*. On the northern slopes of Nebrodi, the Beech forests of the *Anemono apenninae-Fagetum sylvaticae* are replaced by *Taxus baccata* forests (*Ilici aquifolii-Taxetum baccatae*) in stands characterized by colder and more humid and oceanic conditions. On rocky substrata, slightly acidic and well humified of Madonie, Nebrodi and Peloritani, the acidophilous holm oak vegetation named *Geranio versicoloris-Quercetum ilicis* is also found. Within the *Arrhenathero nebrodensis-Quercetum cerridis* distribution area, in the north-facing slopes of the valleys of Peloritani, with remarkably humid microclimatic conditions, the turkey oak vegetation is replaced by *Melitto albidae-Fagetum sylvaticae,* which has to be regarded as an extrazonal community (Brullo et al. 2012).

Fig. 7.3 Monte Soro and its Beech forest during the winter 2012

7.2 Materials and Methods

The main purpose of this work is to compare the actual surface of Beech forest in the Nebrodi Park with the measurements performed by Hofmann (1960). A geographical information system was applied by means of ArcGIS Desktop (Release 9, ESRI, Redlands, CA, US). Different Beech vegetation types were classified on the basis of their density and by field surveys.

The studied area is situated in the Nebrodi Mountains, Sicily's largest mountains complex in terms of surface area, and third in elevation after Mt. Etna and the Madonie. The highest top mountain of Nebrodi Park is Monte Soro (1847 m) (Fig. 7.3).

According to Rivas-Martínez (2004) the bioclimate of this area is Supramediterranean with lower middle-humid bioclimatic conditions (Bazan et al. 2015). On slopes facing north, currents of humid air contribute to the water supply through hidden precipitation, mitigating temperature extremes and frost damage in the spring. At the highest elevations, as in the studied areas, the winter brings heavy snowfall, and the snow sometime remains until the first half of May.

The lithological substrate consists of fissile, blackish or reddish quartzitic sandstone with siliceous cement, of lower Miocene (Aquitanian) age.

In order to outline the real distribution of the Beech forest a set of panchromatic satellite images IKONOS of 2008 with 1 m of resolution, were used. Using a GIS database the measurements performed by Hofmann (1960) were replicated for each

Table 7.1 Variation of the surfaces of Beech forests in the Nebrodi Mountains (1959–2008)

Year 2008			Year 1959	
Municipalities	Surface (hectares)	Inc. (%)	Municipalities	Surface (hectares)
Alcara Li Fusi	818,2007	6.37	Alcara li Fusi	630,0000
Bronte	762,5434	5.93	Bronte	670,0000
Capizzi	987,2671	7.68	Capizzi	1.005,0000
Caronia	1.713,4523	13.33	Caronia	1.440,0000
Cesaro’	4.950,5034	38.51	Cesaro’	4.054,0000
Floresta	137,9648	1.07	Floresta	10,0000
Galati Mamertino	509,3087	3.96	Galati Mamertino	297,0000
Longi	863,2835	6.72	Longi	420,0000
Militello Rosmarino	241,2535	1.88	Militello Rosmarino	340,0000
Mistretta	272,9829	2.12	Mistretta	285,0000
Nicosia	63,5004	0.49	Nicosia	66,0000
Randazzo	21,6840	0.17	Randazzo	18,0000
San Fratello	884,6041	6.88	San Fratello	680,0000
Tortorici	603,7061	4.70	Tortorici	421,0000
Ucria	24,4249	0.19	Ucria	0,0000
Total	12.854,6796	100.00	Total	10.336,0000

territory of municipalities. Using as support the satellite panchromatic cover (edition 2008) and the Regional Map (scale 10.000) the extension of Beech forest of the Nebrodi Park, were measured after about 50 years.

Dendrochronological survey was performed using TSAP Software of Irritech. Regarding climate data, we referred to the worldclim data (Hijmans et al. 2005).

7.3 Results

A total of 12.854 hectares of Beech forest were recognized in the Nebrodi Regional Park distributed in 15 municipalities (Table 7.1) with an increase of 2518 hectares (24.37%) more than 50 years ago (Table 7.2). They are mainly high forest and frequently copse. The average age of high trees is about 90 years (Fig. 7.4). Table 7.2 shows some negative values in some municipalities (e.g. Militello Rosmarino −29.04%). These results are likely imputable to previous mistakes during field measurements, cadastral drafts or confused boundaries of municipalities.

With 4.950 hectares and 38.51% Cesarò is the main core of Beech forest in the Nebrodi Mts followed by Caronia (1.713 ha; 13.33%), Capizzi (987 ha; 7.68%), San Fratello (884 ha; 6.88%); that distribution is more or less comparable with the situation reported 50 years ago. Results are summarized in Fig. 7.5.

Table 7.2 Percentage of variation per municipality during last 50 years

Municipalities	Inc. (%)	Surface variation (hectares)	Surface variation per year (hectares)
Alcara Li Fusi	29.87	188.20	4.71
Bronte	13.81	92.54	2.31
Capizzi	−1.76	−17.73	−0.44
Caronia	18.99	273.45	6.84
Cesaro'	22.11	896.50	22.41
Floresta	1279.65	127.96	3.20
Galati Mamertino	71.48	212.31	5.31
Longi	105.54	443.28	11.08
Militello Rosmarino	−29.04	−98.75	−2.47
Mistretta	−4.22	−12.02	−0.30
Nicosia	−3.79	−2.50	−0.06
Randazzo	20.47	3.68	0.09
San Fratello	30.09	204.60	5.12
Tortorici	43.40	182.71	4.57
Ucria	–	24.42	0.61
Total	24.37	2518.68	62.97

7.4 Discussion

This work allowed to measure the variation of the average of the Beech forests of the Nebrodi Mts equal to 24% with a margin of error of 4%. This increase is mainly due to the change of land use and also thanks to the new Laws for the protection of parks and reserves. In fact, during the second World War, Beech forest of Nebrodi Mts, like other forests in Sicily, was seriously damaged because people took wood for fire and pastoralism was practiced without checks (Fig. 7.6).

It seems indisputable that there isn't influence of climate change, in fact the Fig. 7.7 shows that the rainfall is decreasing during last 50 year.

Currently pasture and alien plants are the main traits affecting the Beech forest of Nebrodi Mts. Inclusion of congeneric *taxa* to the native ones in reforestation projects is the most serious threat of biological pollution and represents an important factor in the genetic erosion (e.g. *Fraxinus excelsior* subsp. *siciliensis* Ilardi & Raimondo). Exotic fauna is another factor dangerous for the Beech renewal. The boar of Balkan and its cross with Sicilian swine represent a very threat for biodiversity conservation like in the main protected areas.

Fagus sylvatica, as a representant of Cenozoic flora in Sicily, forms together *Acer*, *Ulmus*, *Fagus*, *Quercus* a group of woody species that dominate from the level of sea to the high mountains. Other *taxa* like *Pyrus*, *Malus*, *Sorbus* some of which are endemic of Sicily (Marino et al. 2012; Raimondo 2008; Raimondo and Schicchi 2004; Raimondo et al. 2012) grows in theirs shrubby vegetation.

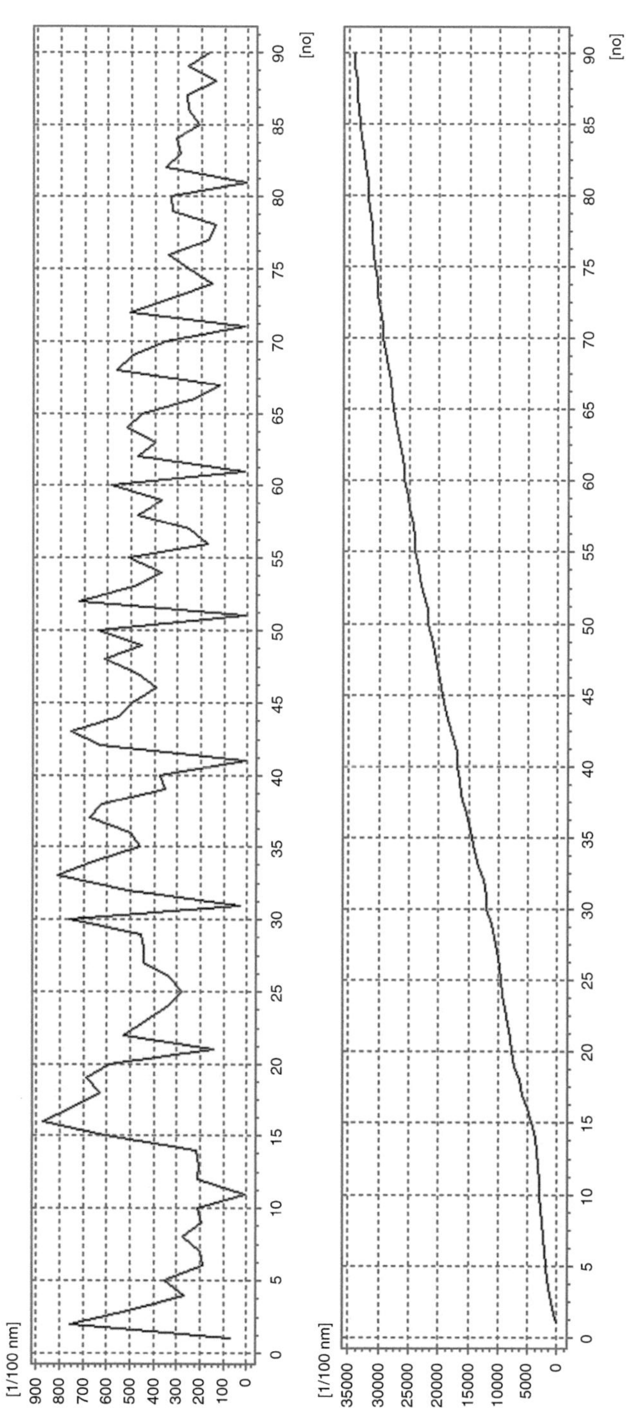

Fig. 7.4 Dendrochronological measurements of some Beech specimens of Nebrodi Mountains

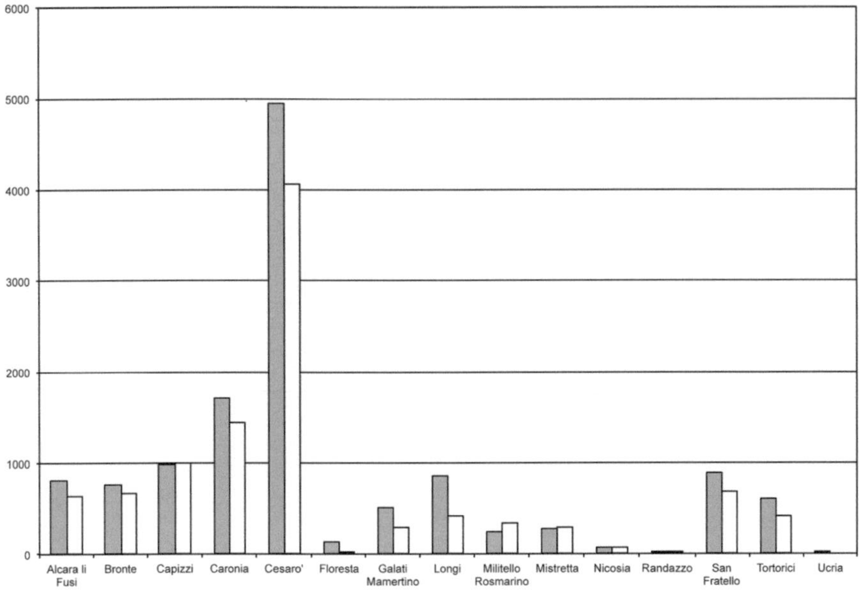

Fig. 7.5 Difference of measurements on Beech forest of Nebrodi Mts in 2008 (*gray*) and in 1959 (*white*)

Fig. 7.6 Footage image where are visible cows and the damaged Beech forest

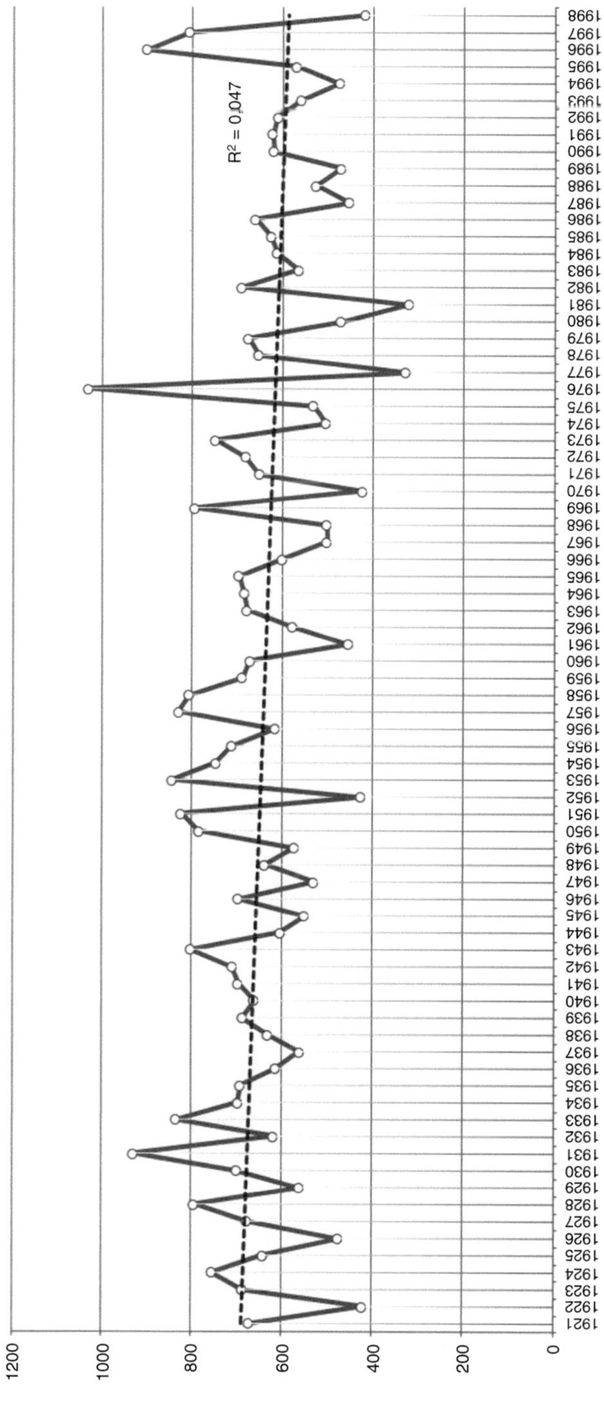

Fig. 7.7 Mean rate of rainfall in Sicily during the last 50 years

Preservation of these habitats is very important for the future of biodiversity. Human activity has shaped these biological resources but tourism especially can result in destructive deterioration. In addition, changing agricultural policies, especially the EU ones, are likely to further change rural landscapes. Conservation strategies of this specific habitat on a regional scale is an overriding objective to preserve biodiversity.

Acknowledgements This work was supported by University of Palermo (Fondi di Ateneo per la Ricerca).

References

Bazan G, Marino P, Guarino R, Domina G, Schicchi R (2015) Bioclimatology and vegetation series in Sicily: a geostatistical approach. Ann Bot Fennici 52(1–2):1–18

Bazan G, Brullo S, Raimondo FM, Schicchi R (2010) Le serie di vegetazione della Regione Sicilia. In: Blasi (ed) La vegetazione d'Italia. Palombi Editori, Roma, pp 429–470

Brullo C, Brullo S, Giusso del Galdo G, Guarino R, Siracusa G, Sciandrello S (2012) The class *Querco-fagetea sylvaticae* in Sicily: an example of boreo-temperate vegetation in the central mediterranean region. Ann Bot (Roma) 2:19–38

Del Favero R (2008) I boschi delle regioni meridionali e insulari d'Italia. Tipologia, funzionamento, selvicoltura. CLEUP, Padova

Ellemberg H (1996) Vegetation Mittleuropas mit den Alpen. Ulmer, Stuttgart

Fenaoli L (1970) Note illustrative della carta della vegetazione reale d'Italia. Ministero Agricoltura e Foreste, pp 96–98

Hijmans RJ, Cameron SE, Parra JL, Jone PG, Jarvis A (2005) Very high resolution interpolated climate surfaces for global land areas. Int J Climatol 25:1965–1978

Hofmann A (1960) Il faggio in Sicilia. Gianasso, Sondrio

Hofmann A (1961) La faggeta depressa del Gargano. Delpinoa 3:375–406

Marino P, Castellano G, Raimondo FM, Spadaro V (2012) Pyrus ciancioi (Rosaceae) a new species from Sicily. Plant Biosyst 146(3):654–657

Nocentini S (2009) Structure and management of beech (*Fagus sylvatica* L.) forests in Italy. iForest 2:105–113

Palamarev E, Mai DH (1998) Die paläogenen Fagaceae in Europa: Artenvielfalt und eitlinienihrer Entwicklungsgeschichte. Acta Palaeobot 38(2):227–299

Pignatti S (1998) I boschi d'Italia. UTET, Torino

Raimondo FM (2008) Malus crescimannoi (Rosaceae) a new species from Sicily. Fl Medit 18:5–10

Raimondo FM, Schicchi R (2004) Pyrus vallis-demonii (Rosaceae) a new species from the Nebrodi Mountains (NE-Sicily). Bocconea 17:325–330

Raimondo FM, Castellano G, Bazan G, Schicchi R (2012) Sorbus madoniensis (Rosaceae), a new species from Sicily. Plant Biosyst 146(1):345–351

Rivas-Martínez S (2004) Global Bioclimatics (Clasificación Bioclimática de la Tierra). Available at http://www.globalbioclimatics.org. Accessed March 2014

The Magnocaricetalia Pignatti 1953 (*Phragmito-Magnocaricetea* Klika in Klika et Novák 1941) Plant Communities of Italy

8

Roberto Venanzoni, Alessandro Properzi, Emma Bricchi, Flavia Landucci, and Daniela Gigante

Abstract

The Authors present a list of marshy and wet vegetation described or present in Italy included in the *Magnocaricetalia* Pignati 1953 Order (*Phragmito-Magnocaricetea* Klika in Klika et Novák 1941 Class) focusing on nomenclatural, floristic, physiognomic and ecological traits reported by Authors in the examined literature.

Keywords

Magnocaricetalia • Plant communities • Italy

Marshes and aquatic vegetation play a vital role in the global hydrological cycle delivering many important services to the ecosystems, such as providing food, increasing the transparency of the water transporting away and filtering wastewater, controlling groundwater, floods and soil erosion, (Beaufoy et al. 1994; Postel and Carpenter 1997; Bignal and McCraken 2000; IUCN 2008), preserving biodiversity both in agroecosystems and natural landscapes (Hooper et al. 2005; Taffetani et al. 2011; ISPRA 2011). It is largely distributed in Central Europe and extends its presence to all European Biogeographic Regions (Fig. 8.1). This azonal vegetation type is mainly primary but it is also abundant as secondary vegetation in substitution of riparian and swampy forests after their cutting (Balátová-Tuláčková 1963, 1994; Balátová-Tuláčková and Pavlíček 1980; Balátová-Tuláčková et al. 1993; Oberdorfer 1977; Pott and Remy 2000).

According to the Phytosociolocal syntaxonomic point of view marshy vegetation is included into the Class *Phragmito-Magnocaricetea,* which includes five

R. Venanzoni (✉) • A. Properzi • E. Bricchi • F. Landucci • D. Gigante
Dipartimento di Chimica, Biologia e Biotecnologie, Università di Perugia, Via Elce di Sotto 8, 06123 Perugia, Italy
e-mail: roberto.venanzoni@unipg.it

F. Pedrotti (ed.), *Climate Gradients and Biodiversity in Mountains of Italy*,
Geobotany Studies, https://doi.org/10.1007/978-3-319-67967-9_8

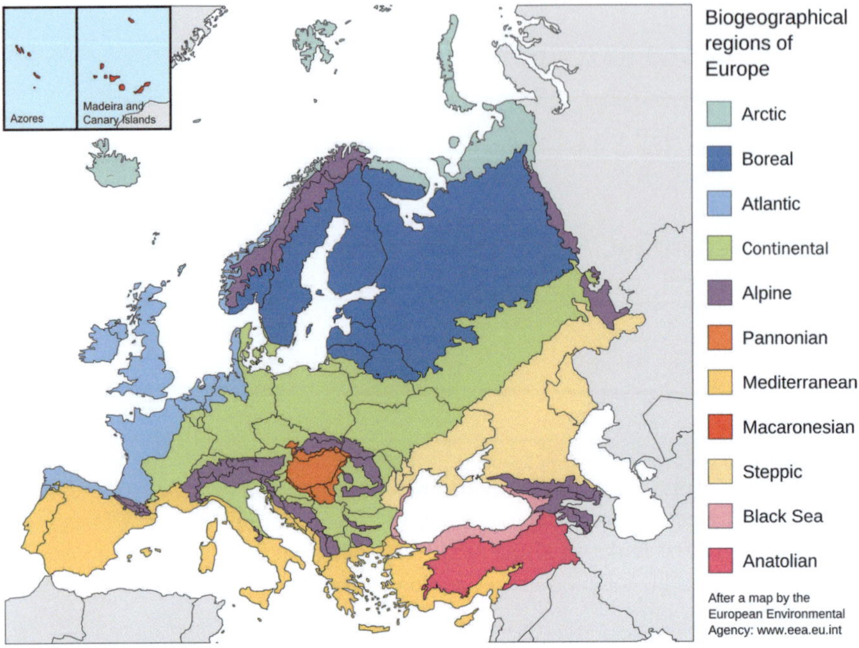

Fig. 8.1 Biogeographic subdivision of Italy according to UE Biogeographic subdivision

Orders: *Phragmitetalia* Kock 1926, *Magnocaricetalia* Pignatti 1953, *Scirpetalia compacti* Heijný in Holub, Heijný, Moravec & Neuhäusl 1967 corr. Rivas-Martínez, Costa, Castroviejo & E. Valdés 1980 (syn. *Bolboschoenetalia maritimi* Hejny in Holub et al. 1967), *Nasturtio-Glyceretalia* Pignatti 1953, *Oenanthethalia aquaticae* Hejný in Kopecký & Hejný 1965 (Balátová-Tuláčková et al. 1993; Pott and Remy 2000; Oťahelová et al. 2001; Biondi et al. 2014).

Since 1964, when Phytosociology was formally developed in Italy with the constitution of the Italian Society of Phytosociology (see Not. Fitosoc. n.1), a great amount of literature has been published including the Pignatti's milestone monograph in which he described for the first time the Order *Magnocaricetalia* and other syntaxa of the same Class (Pignatti 1953).

According to a recent study on the class *Phragmito-Magnocaricetea* (Landucci et al. 2013) 43 associations are present in Central Italy. Thirteen of them belong to *Magnocaricetalia* Order, demonstrating a great diversity of communities similar to what observed in Central Europe (Matulevičiūtė 2002) and Iberian Peninsula (Molina and Moreno 2003). This pattern confirms that Italy during the last glaciations was an important refuge for the genus *Carex* (Jiménez-Mejías et al. 2011, 2014). Sedge-dominated communities, in most cases, represent a relict vegetation type related to soil condition originated in the past post-glacial climate conditions presently no longer available at the same latitude (Pedrotti and Sanesi 1969; Cortini et al. 1973).

In this paper we present a list of marshy and wet vegetation syntaxa reported during the last 60 years in Italy, mainly included in the *Magnocaricetalia* Order according to different and not always homogeneous authors interpretations. In order to offer an updated overview of the syntaxa recorded without any effort to solve the many syntaxonomical critical issues still under discussion and to enforce the conservation status.

8.1 Material and Methods

For this work all the syntaxonomic scientific literature produced in Italy present in the database Lisy "www.scienzadellavegetazione.it" (Bracco and Nola 1995; Biondi et al. 1996, 1997; Bracco 2001; Bracco et al. 2007) and forward publications was examined. For each paper phytosociological relevés were extracted, georeferenced and stored in the database VegItaly (www.vegitaly.it) the national vegetation database project of the Italian Society of Vegetation Science (Gigante et al. 2012; Landucci et al. 2012; Venanzoni et al. 2012). Species taxonomy and nomenclature were unified according to the checklist of the Italian vascular plants (Conti et al. 2005) and the work of synonymization made in the global botanical database www.anarchive.it project (Lucarini et al. 2015).

Most important syntaxonomical European literature has been compared during the formal work (Oberdorfer 1977; Grabherr and Mucina 1993; Valachovič 2001; Schaminée et al. 1995; Ninot et al. 2000; Rivas-Martínez et al. 2001, 2002; Bardat et al. 2001, 2004; Chytrý 2011; Biondi et al. 2014) to have a formal syntaxonomical prospect of the Italian vegetation syntaxa according to the International Code of Phytosociological Nomenclature (Weber et al. 2000).

8.2 Results

According to the investigated literature *Magnocaricetalia* Order in Italy includes 4 alliances, 55 communities and 22 sub-units reported in the following syntaxonomic scheme:

PHRAGMITO AUSTRALIS-MAGNOCARICETEA ELATAE Klika in Klika et Novák 1941
Magnocaricetalia elatae Pignatti 1953
[Syn. *Magnocaricetalia* Pignatti 1953 1954, *Magnocaricetalia* Pignatti 1954]

***Magnocaricion elatae* Koch 1926**

[Syn. *Caricion gracilis* Neuhäeusel 1959, *Caricion rostratae* Neuhäeusel 1959, *Caricion appropinquatae* Balátová-Tuláčková 1960, *Caricion rostratae* Balátová-Tuláčková 1963]

- *Caricetum elatae* Koch 1926
  [Syn: *Scutellario-Caricetum elatae* Passarge 1964]
  typicum Jeschke 1955

comaretosum palustris Passarge 1955
uticularietosum vulgaris Jeschke 1955
phragmitetosum australis Gerdol 1988 n.n.
caricetosum hartmanni Balátová-Tuláčková 1976
hydrocotyledonetosum vulgaris Vagge et Biondi 1999
variant with *Carex acutiformis*
variant with *Eupatorium cannabinum*
facies with *Carex appropinquata*

- *Carici elatae-Calamagrostietum canescentis* Jílek 1958
  [Syn: *Peucedano-Calamagrostietum canescentis* Weber 1978, *Cirsio palustris-Calamagrostietum canescentis* Passarge 1984, *Calamagrostietum canescentis* Simon 1960 *calamagrostietosum canescentis* Hadač in Březina et al. 1963]
- *Filipendulo ulmariae-Caricetum randalpinae* Stančić 2009
- *Caricetum acutiformis* Eggler 1933
  [Syn: *Caricetum acutiformis* Sauer 1937, *Caricetum acutiformi-ripariae* Soó 1938, *Caricetum acutiformis* Sauer ex Weber 1977, *Caricetum ripariae-acutiformis* Kobenza 1930; incl. *Galio palustris-Caricetum acutiformisAutori? Galio elongati-Caricetum acutiformis* Molina et Curcó in Curcó 2000]
- *Caricetum appropinquatae* Aszód 1935
  [Syn: *Caricetum paradoxae* Aszód 1936, *Caricetum diandro-appropinquate* Klika 1958]
- *Caricetum cespitosae* Dagys 1932
  [Syn: *Caricetum cespitose* Steffen 1931]
- *Caricetum diandrae* Jonas 1933
- *Peucedano palustris-Caricetum lasiocarpae* Tüxen ex Balátová-Tulácková 1972
- *Galio palustris-Caricetum rostratae* Martinčič et Seliiškar 2004
  [Syn: *Caricetum rostratae* (Dagys 1932) Balátová-Tulácková 1963, *Caricetum rostratae* Osvald 1923, *Equiseto fluviatilis-Caricetum rostratae* Zumpfe 1929, *Caricetum rostratae* Rübel 1912]
- *Thelypterido palustris-Phragmitetum australis* Kuiper ex van Donselaar 1961
- *Cladietum marisci* Allorge 1921
  [Syn. *Phragmito-Cladietum marisci* von Soó 1930, *Mariscetum serrati* Zobrist 1935]
  typicum Pfeifer 1961
  schoenetosum nigricantis Dierßen 1982
  juncetosum subnodulosi Dierßen 1982
- *Junco maritimi-Cladietum marisci* Géhu *et* Biondi 1989
- *Soncho maritimi-Cladietum marisci* (Braun-Blanq. *et* O. Bolòs 1958) Cirujano 1980
  [Syn. *Mariscetum oligohalinum* Braun-Blanq. *et* O. Bolòs 1957]
  myrtetosum communis Biondi, Casavecchia et Guerra 2006
- *Leucojo-Caricetum elatae* Braun-Blanq. 1936

[Incl. *Leucojo-Caricetum elatae* Braun-Blanq. 1936 *caricetosum elatae*]
- *Leucojo-Caricetum ripariae* Braun-Blanq. (1931) 1936
  [Incl. *Leucojo-Caricetum elatae* Braun-Blanq. 1936 *caricetosum ripariae*]
- *Eupatorio cannabini-Caricetum elatae* Biurrun, J.A. Molina *et* Loidi in Biurrun 1999
- *Eleocharitetum palustris* Savič 1926
  [Syn. *Eleocharis palustris* community Schennikov 1919, *Eleocharidetum palustris* Schennikow 1919, *Eleocharitetum palustris* Ubrizsy 1948, *Eleocharidetum palustris* Lakušić et Pavlović 1976; *Eleocharidetum palustris* Ubrizsy 1948; *Eleocharidetum palustris* Schennikov 1919 ex Ubrizsy 1948] [inl. phytocoenon a *Eleocharis palustris*]
  paspaletosum paspaloidis Biondi et al. 2013 ex Biondi, Casavecchia et Radetic 2002
- *Eleocharitetum uniglumis* Almquist 1929
  [incl. *Eleocharis uniglumis* community]
- *Caricetum hispidae* Brullo et Ronsisvalle 1975
  [syn. *Cladio-Caricetum hispidae* O. Bolós 1967 *p.p.*]
- *Caricetum otrubae* Mirza 1978
  [non *Caricetum otrubae* Pedrotti 1982]
- *Cypero-Caricetum cuprinae* Tüxen ex T.E. Díaz & F. Prieto 1994
  [Syn. *Cypero longi-Caricetum otrubae* R. Tüxen. in R. Tüxen. *et* Oberd. 1958]
  typicum
  euphorbietosum pubescentis O. Bolòs 1962
- *Cyperetum longi* Micevski 1957
- *Carici distantis-Schoenetum nigrescentis* Brullo, Minissale, Scelsi et Spampinato 1993
- *Leersio oryzoidis-Juncetum effusi* Lastrucci, Paci et Raffaelli 2010
  [incl. *Leersietum oryzoidis* Eggler 1933, phytocenon with *Leersia oryzoides*]
- *Galio palustris-Juncetum inflexi* Venanzoni *et* Gigante 2000
- *Equiseto palustris-Juncetum effusi* Minissale *et* Spampinato 1990
- *Mentho aquaticae-Typhetum minimae* Venanzoni *et* Gigante 2000
- *Hyperico tetrapteri-Caricetum acutiformis* Corbetta *et* Pirone 1990
- *Holoschoeno-Juncetum subnodulosi* Géhu *et* Biondi 1988
- *Iridetum pseudacori* Eggler ex Brzeg et M. Wojterska 2001
- *Caricetum pendulo-panormitanae* Gianguzzi et al. 2013
- *Mentha aquatica* community
  [Incl. *Mentha aquatica phytocoenon,* groupment with *Lycopus europaeus and Mentha aquatica*]
- *Ranunculus flammula* and *Juncus articulatus* community
- Groupment with *Menyanthes trifoliata*
- Groupment with *Juncus subnodulosus*
- *Teucrio scordii-Juncetum subnodulosi* Pellizzari & Piccoli 2001
  utricularietosum autralis Pellizzari & Piccoli 2001
- Groupment with *Juncus effusus* and *Galium elongatum*

- Groupment with *Lysimachia vulgaris*
- Groupment with *Equisetum fluviatile*
- Groupment with *Veronica beccabunga*
- *Veronica anagallis-aquatica Phytocoenon*

***Caricion gracilis* Géhu 1961**

[Syn. *Magnocaricion elatae* Koch 1926 p.p., *Caricion gracilis-vulpinae* Balátová-Tuláčková 1965, *Caricenion gracilis* (Neuhäusl 1959) Oberdorfer et al. 1967]

- *Caricetum gracilis* Savič 1926
  [Syn. *Caricetum acuto-vesicariae* (Koch 1926) Westhoff 1949, *Caricetum acutiformi-gracilis* von Soò 1927, *Caricetum gracilis* Graebner et Hueck 1931, *Caricetum gracilis* Eggler 1933, *Caricetum gracilis* (Almquist 1929) Tüxen 1937] [Incl. Phytocoenon with Cares gracilis, groupment with *Carex riparia* and *Carex gracilis*]
  typicum Cortini et al. 1973
  caricetosum buxbaumi Cortini et al. 1973
- *Caricetum distichae* Nowinski 1927
  [Syn. *Caricetum intermediae* Steffen 1931, *Caricetum distichae* Jonas 1933]
- *Caricetum buxbaumii* Issler 1932
- *Ranunculo apennini-Caricetum buxbaumi* Pedrotti 2016
  [Syn. Ranunculo-Caricetum buxbaumi Pedrotti 2015 *nomen invalidum*]
- *Caricetum paniculatae* Wangerin 1916
  [Syn. *Caricetum acutiformi-paniculatae* Vlieger et van Zinderen Bakker in Boer 1942, *Caricetum paniculatae* Wangerin ex von Rochow 1951, *Eupatorio-Caricetum paniculatae* Passarge 1999]
- *Caricetum ripariae* Máthé et Kovács 1959
  [Syn. *Caricetum ripariae* Soó 1928, *Carex riparia*-Gesellschaft Knapp et Stoffers 1962, *Galio palustris-Caricetum ripariae* Balátová-Tuláčková in Balátová-Tuláčková et al. 1993] [Incl. groupment with *Carex riparia* and *Carex gracilis*]
- *Caricetum vesicariae* Chouard 1924
  [Syn. *Caricetum vesicariae* Braun-Blanquet et Denis 1926, *Caricetum inflato-vesicariae* Koch 1926, *Caricetum acuto-vesicariae* (Koch 1926) Westhoff 1949 p.p., *Caricetum vesicariae* Rübel 1933, *Caricetum vesicariae* Eggler 1933]
- *Caricetum vulpinae* Nowiński 1927
  [Syn. *Caricetum vulpinae* von Soó 1927; *Caricetum otrubae* Pedrotti 1982]
- *Carici hispidae-Juncetum subnodulosi* Gehu et Biondi 1989
- *Phalaridetum arundinaceae* Libbert 1931
- Groupment with *Carex melanostachya*

***Carici-Rumicion hydrolapathi* Passarge 1964**

[*Oenanthion aquaticae* Hejný ex Neuhäusl 1959, *Cicution virosae* Hejný ex Segal in Westhoff et den Held 1969]

- Cicuto virosae-Caricetum pseudocyperi Boer et Sissingh in Boer 1942
  [Syn. *Caricetum pseudocyperi* Boer 1942]
  caricetosum paniculatae Tomaselli et al. 2006
- *Mentho aquaticae-Caricetum pseudocyperi* Orsomando et Pedrotti 1986

***Caricion microcarpae* Gamisans (1968) 1975**

- *Hyperico hircini-Caricetum microcarpae* Bacchetta et Mossa 2004
  [*Hypertico hircini-Caricetum microcarpae* Bacchetta e Mossa 2002 n.n., *Hypertico hircini-Caricetum microcarpae* 2008 in Mossa et al. 2008]
  oenanthetosum crocatae Bacchetta et Mossa 2004
  adiantetosum capilli-veneris Bacchetta et Mossa 2004

8.3 Discussion and Conclusions

Italy presents a large number of communities: 54 with many sub-units mainly in central-northern part with a decreasing gradient from North to South.

The *Magnocaricion elatae* alliance is largely consisting of tall sedges-communities that grow on organic and hydromorphic soils often peaty and flooded, from mesotrophic to dystrophic habitat conditions. It includes about 80% of the described communities and for that reason seems very uneven. In Italy it has an inland distribution but goes up to the more South-Mediterranean regions, including major Islands (Sicily and Sardinia) with a very reduced set of associations. At the same time along the North-South gradient we observe a reduction of typical floristic elements belonging to the central European floristic Region, without a corresponding significant increase of endemic Mediterranean species and. Instead, in most cases, in the South the communities are enriched of species belonging to the Orders *Molinio-Schoenetalia*, *Phragmitetalia*, *Scirpetalia compacti*, *Juncetalia maritimae* that contribute to characterize these marshy communities of *Magnocaricion* at the edge of the their distribution area (Brullo and Ronsisvalle 1975; Brullo et al. 1993; Venanzoni and Gigante 2000). In this context two of these endemic species are rapresented by *Carex panormitana* and *C. microcarpa*.

In the alpine Region many associations of this alliance characterized by *Carex rostrata*, *C. diandra* and *C. lasiocarpa* are closer to peaty vegetation of mires and bogs of *Scheuchzerio-Caricetea fuscae* and *Oxycocco-Sphagnetea* syntaxa depending to the hight presence of mosses and *Sphagnum* species.

The *Caricion elatae* alliance could be divided into two sub-alliances the first one for Central European communities and the second one, less populated by continental tall sedges and characteristic species of higher units, for associations with sub-Mediterranean and Mediterranean distribution (Table 8.1) according to European biogeographic map in Fig. 8.1.

Table 8.1 Biogeographic range of the *Magnocaricetalia* communities according of UE Biogeographic subdivision (AC = Alpine & Continental, CM = Continental & Mediterranean, M = Mediterranean), corresponding literature of each syntaxa (extended citation of literature is reported in Annex 1) not published releves are mostly available on the website database www.vegitaly.it

	Communities	AC	CM	M	Literature
	Magnocaricion elatae				
1	*Caricetum elatae*	X	X	–	Pignatti (1953); Andreis et al. (1995b); Andreis and Zavagno (1996); Andrich et al. (2001); Badino et al. (1982–1983); Bertolani Marchetti et al. (1977); Bracco (1981); Bracco and Nola (1995b); Bracco et al. (2000); Brusa et al. (2006a, b); Buchwald (1994); Canullo et al. (1991); Canullo et al. (1994); Corbetta et al. (1998); Focarile (1977); Gansi (2007–2008); Gerdol and Tomaselli (1997); Gerdol (1997); Guglielmetto et al. (1993–1994); Lasen and Argenti (1996); Lastrucci et al. (2008); Lonati and Lonati (2005); Marchiori et al. (1980, 1993); Minghetti and Pedrotti (2000); Pedrotti and Chemini (1981); Pedrotti and Cortini Pedrotti (1996); Pedrotti and Gafta (1992); Pedrotti and Venanzoni (1987); Pedrotti (1967, 1980, 1986, 1983, 1987, 1988b, 1989, 1990, 1991, 1995, 2000, 2001, 2003, 2004); Pedrotti et al. (1992); Pellizzari et al. (2005); Piccoli et al. (1979); Piccoli (1976); Pirola (1968); Pirone and Corbetta (1998); Poldini (1989); Prosser and Sarzo (2003, 2004); Rey (1990); Sarzo et al. (1999); Sburlino and Marchiori (1985); Sburlino et al. (1995); Scortegagna and Tasinazzo (2002); Tammaro et al. (1986); Tasinazzo (2009); Tisi et al. (2007); Tomaselli et al. (2006); Venanzoni and Gigante (2000); Gigante e Venanzoni (2004); Venanzoni (1988b, 1995); Vagge and Biondi (1999); Boiti et al. (1990); Corsi (2001); Bolpagni et al. (2007)
	typicum	–	X	–	Balátová-Tuláčková and Venanzoni (1989)
	phragmitetosum australis [incl. var with *Phragmites australis*]	–	X	–	Gerdol (1987); Tomaselli et al. (2006); Pedrotti (1986) come sopra
	comaretosum palustris	–	X	–	Balátová-Tuláčková and Venanzoni (1989)
	uticularietosum vulgaris	–	X	–	Balátová-Tuláčková and Venanzoni (1989)
	caricetosum hartmannii	–	X	–	Pedrotti (2004)
	hydrocotyledonetosum vulgaris	–	X	–	Vagge and Biondi (1999)
	variant with *Carex acutiformis*	–	X	–	Tomaselli et al. (2006)

	variant with *Eupatorium cannabinum*	–	X	–	Tomaselli et al. (2006)
	facies with *Carex appropinquata*	X	–	–	Rey (1990)
2	*Carici elatae-Calamagrostietum canescentis* [Incl. *Calamagrostietum canescentis*]	X	–	–	Guglielmetto et al. (1993–1994); Gerdol (1987); Sartori and Bracco (1997); Tomaselli et al. (2006); Brusa et al. (2006b)
	Filipendulo ulmariae-Caricetum randalpinae Stančić 2009	X	–	–	Lasen (2014) *in verbis*
3	*Leucojo-Caricetum elatae* [Incl. *Leucojo-Caricetum elatae caricetosum elatae*]	–	X	–	Pirola (1968); Piccoli et al. (1979); Merloni and Piccoli (2001, 2007); Assini et al. (2010); Sburlino (2014) in verbis
4	*Leucojo-Caricetum ripariae* [incl. *Leucojo-Caricetum elatae caricetosum ripariae*]	–	X	–	Pirola (1968)
5	*Eupatorio cannabini-Caricetum elatae*	–	X	–	Biondi et al. (2002b); Lastrucci et al. (2012); Gubellini et al. (2014)
6	*Caricetum acutiformis* [Incl. groupement with *Carex acutiformis*, *Carex acutiformis* community]	–	X	–	Andreis et al. (1993); Assini et al. (2010); Badino et al. (1982–1983); Balátová-Tuláčková and Venanzoni (1989); Biondi et al. (1997, 1999); Brusa et al. (2006a, b); Corbetta and Pirone (1989); Pirone e Corbetta (1998); Corbetta et al. (1998); Gansi (2007–2008); Gerdol (1987); Giau (1986); Giglio and Tammaro (1995); Lastrucci et al. (2008); Luchetta (2001); Pedrotti and Cortini Pedrotti (1996); Pedrotti and Gafta (1994); Pedrotti (1995, 2008); Pellizzari et al. (2005, 2007); Piccoli et al. (1996, 1999); Pirone et al. (1997); Prosser and Sarzo (2003, 2004); Sarzo et al. (1999); Scortegagna and Tasinazzo (2000, 2001, 2002); Venanzoni (1988c, 1992, 1995); Venanzoni and Gigante (2000); Gigante and Venanzoni (2004); Venti et al. (2003); Zanotti et al. (1981); Taffetani et al. (2011)
7	*Caricetum appropinquatae*	X	–	–	Pedrotti (1995); Venanzoni (1995); Balátová-Tuláčková and Venanzoni (1989); Brusa et al. (2006b); Guglielmetto Mugion and Montacchini (1993–1994)
8	*Caricetum caespitosae*	X	–	–	Montanari and Guido (1980); Badino et al. (1982); Conti et al. (2005)
9	*Caricetum diandrae*	X	–	–	Rey and Desfayes (1989); Rey (1990); Pedrotti (1991); Venanzoni (1984); Balátová-Tuláčková and Venanzoni (1989); Pedrotti (1988 [estinct], 1991); Lasen and Argenti (1996); Andrich et al. (2001)

(continued)

Table 8.1 (continued)

	Communities	AC	CM	M	Literature
10	*Peucedano palustris-Caricetum lasiocarpae*	X	–	–	Rey and Desfayes (1989); Rey (1990); Bonomi and Buffa (2000); Pedrotti (1980, 1981–82, 1984, 1985, 1990, 2003, 2001, 2004); Pedrotti and Chemini (1981); Canullo et.al. (1991, 1994); Luchetta (2001); Andreis e Rotondi (1982); Cristea et al. (2004)
11	*Galio palustris-Caricetum rostratae* [Incl. groupement with *Carex rostrata*]	X	X	–	Rey (1990); Venanzoni (1988b); Ferrari e Manzi (1987); Tisi et al. (2007); Badino et al. (1982); Venanzoni (1984); Pedrotti et al. (1982, 1985); Venanzoni Lago Pudro [not published]; Pedrotti (1969, 1974, 1980, 1981, 1991, 2001, 2003, 2004); Prosser e Sarzo (2003, 2004); Sburlino (1986); Balátová-Tuláčková and Venanzoni (1990); Pedrotti and Venanzoni (1987); Bonomi and Buffa (2000); Canullo et al. (1990); Lasén and Argenti (1996); Caniglia et al. (1992); Andrich et al. (2001); Luchetta (2001); Lausi and Gérdol (1980); Aita et al. (1976); Montanari and Guido (1980); Gentile et al. (1988); Gerdol e Bragazza (2001); Andreis and Rodondi (1982); Foggi et al. (2007); Gargano et al. (2007); Gerdol and Tomaselli (1997); Canullo et al. (1994); Pedrotti (1981, 1991); Gerdol and Tomaselli (1993); Argenti et al. (2006); Minghetti e Pedrotti (2000)
12	*Thelypterido palustris-Phragmitetum australis*	–	X	–	Pedrotti (1988, 1990, 1991); Venanzoni (1995); Venanzoni (2014); Lasen (2014), *in verbis*; Andreis et al. (1993); Arrigoni (1982)
13	*Cladietum marisci*	–	X	–	Focarile (1977); Pedrotti (2001); Pedrotti (1980, 1988); Arrigoni and Riccieri (1981); Giau (1986); Biondi (1986) Gerdol (1987); Gèhu e Biondi (1988); Poldini (1989); Balátová-Tuláčková and Venanzoni (1989); Dell'Uomo and Agostinelli (1990); Mariotti et al. (1992); Andreis et al. (1993); Pedrotti and Minghetti (1994); Andreis and Zavagno (1996); Biondi et al. (1997); Bracco et al. (1998); Sarzo et al. (1999); Venanzoni and Gigante (2000); Merloni and Piccoli (2001, 2007); Landi and Angiolini (2006); Pirone and Corbetta (1998); Tomei et al. (1998); Venanzoni et al. (2003); Brusa et al. (2006b); Biondi et al. (2007); Tisi et al. (2007); Giugliano e Terzani (2011); Studio verde (2013); Sburlino et al. (2013)
	typicum	–	X	–	Balátová-Tuláčková and Venanzoni (1989)
	schoenetosum nigricantis	–	X	–	Balátová-Tuláčková and Venanzoni (1989)
	juncetosum subnodulosi	–	X	–	Balátová-Tuláčková and Venanzoni (1989)

14	*Junco maritimi-Cladietum marisci*	–	–	X	Gèhu and Biondi (1988, 1996)
15	*Soncho maritimi-Cladietum marisci*	–	–	X	Brullo e Ronsisvalle (1975); Brullo e Furnari (1976); Bartolo et al. (1982); Arrigoni et al. (1985); Giau (1986) p.p.; Arrigoni (1990); Mariotti et al. (1992); Brullo et al. (1993); Buchwald (1994); Biondi et al. (1997); Vagge e Biondi (1999); Tomaselli (2001); Brullo et al. (2002); Tomaselli et al. (2010, 2011); Reniero et al. (2013); Sburlino et al. (2013)
	myrtetosum communis	–	–	X	Biondi et al. (2006)
16	*Caricetum hispidae*	–	X	X	Brullo and Ronsisvalle (1975); Brullo (1978); Gèhu and Biondi (1983); Bartolo et al. (1982); Brullo (1980); Brullo et al. (1996, 2002); Raimondo et al. (2000); Brullo and Spampinato (1990); Brullo and Furnari (1976); Sortino et al. (1974)
17	*Caricetum otrubae*	–	X	–	Pirone (1987); Corbetta and Pirone (1989); Buchwald (1994); Brullo and Sciandrello (2006); Angiolini et al. (2006); Lastrucci et al. (2010a); Lasen (2014) *in verbis*
18	*Cypero-Caricetum cuprinae*	–	X	X	Barbagallo et al. (1979b); Brullo and Spampinato (1990); Minissale and Spampinato (1995); Venanzoni (1998); Venanzoni and Gigante (2000); Raimondo et al. (2000); Tomaselli (2001); Brullo et al. (2002); Gasparri et al. (2013); Venanzoni (2013); Venanzoni [not published releves]
	typicum	–	X	X	Venanzoni [not published releves]
	euphorbietosum pubescentis	–	X	X	Venanzoni [not published releves]
19	*Carici distantis-Schoenetum nigrescentis*	–	–	X	Brullo et al. (1993); Brullo et al. (2002); Tomaselli (2001, 2004)
20	*Cyperetum longi*	–	X	–	Barbagallo et al. (1979a); Corbetta e Pirone (1989); Brullo and Spampinato (1990); Andreis et al. (1993); Brullo et al. (1994, 1996, 2002); Tammaro (1995); Pirone et al. (1997, 2003); Pirone e Corbetta (1998); Raimondo et al. (2000); Venanzoni and Gigante (2000); Tomaselli (2001); Venti et al. (2003); Venanzoni et al. (2003); Pedrotti (2008); Cianfaglione (2009) (2012); Lastrucci et al. (2010a, b, 2012); Sburlino (2014) *in verbis*; Lasen (2014) *in verbis*. Poli e Maugeri (1974)
	typicum	–	X	–	Barbagallo et al. (1979a)
	caricetosum pendulae	–	X	X	Barbagallo et al. (1979a)
21	*Leersio oryzoidis-Juncetum effusi*	–	X	–	Lastrucci et al. (2010); Scoppola (1998); Poldini (1989)

(continued)

Table 8.1 (continued)

	Communities	AC	CM	M	Literature
22	*Equiseto palustris-Juncetum effusi*	–	X	–	Minissale and Spampinato (1990); Venanzoni and Gigante (2000); Lasen (2014) *in verbis*
24	*Mentha aquatica* community	–	X	–	Barberis and Mariotti (1981); Gansi (2007–2008); Lastrucci et al. (2010a); Lasen (2014) *in verbis*; Sburlino (2014) *in verbis*
25	*Mentho aquaticae-Typhetum minimae*	–	X	–	Venanzoni and Gigante (2000)
26	*Holoschoeno-Juncetum subnodulosi*	–	X	–	Gèhu and Biondi (1988); Merloni and Piccoli (2007)
27	*Iridetum pseudacori*	–	X	–	Pedrotti (1995); Prosser and Sarzo (2003); Pellizzari et al. (2005); Ceschin and Salerno (2008); Lastrucci et al. (2008, 2010b); Tomaselli et al. (2011); Vena zoni (2014) *in verbis*
28	*Galio palustris-Juncetum inflexi* [Incl. groupement with *J. inflexus*]	–	X	X	Corbetta and Pirone (1989); Venanzoni and Gigante (2000, 2004); Venti et al. (2003); Ceschin and Salerno (2008); Lastrucci et al. (2012); Lasen (2014) *in verbis*.
29	*Hyperico tetrapteri-Caricetum acutiformis*	–	X	–	Corbetta and Pirone (1990); Tammaro (1995)
30	*Ranunculus flammula - Juncus articulatus community*	–	X	–	Gargano et al. (2007); Lasen (2014) *in verbis*
31	Groupement with *Menyanthes trifoliata*	X	–	–	Montanari and Guido (1980); Andreis and Rodondi (1982); Gerdol and Tomaselli (1993); Venanzoni (1995); Gargano et al. (2007); Foggi et al. (2007); Sburlino (2014) *in verbis*; Lasen (2014) *in verbis*
32	Groupment with *Juncus subnodulosus*	–	X	–	Pedrotti et al. (1979); Balátová-Tuláčková e Venanzoni (1989); Piccoli and Merloni (1989); Venanzoni and Gigante (2000); Prosser and Sarzo (2004)
33	*Teucrio scordii-Juncetum subnodulosi*	–	–	X	Pellizzari e Piccoli (2001)
	utricularietosum autralis	–	–	X	Pellizzari e Piccoli (2001)
34	Groupment with *Juncus effusus* e *Galium elongatum*	–	X	–	Lastrucci et al. (2007)
35	Groupement with *Lysimachia vulgaris*	–	X	–	Venanzoni (1995); Lastrucci et al. (2007)
36	Groupment with *Equisetum fluviatile*	–	X	–	Gerdol and Tomaselli (1993)
37	Groupment with *Veronica beccabunga*	X	–	–	Vanacore Falco and Venanzoni (2007–08)

38	*Eleocharitetum uniglumis* [Incl. *Eleocharis uniglumis* community]	–	X	–	Braggio Morucchio et al. (1979); Piccoli and Merloni (1989); Merloni and Piccoli (2007); Lasen (2014) *in verbis*
39	*Eleocharitetum palustris* Schenn. 1919 [Incl. groupment with *E. palustris*, phytocoenon with *E. palustris*]	–	X	–	Poldini (1989); Rey (1990); Pedrotti and Gafta (1992); Biondi et al. (1997); Venanzoni e Gigante (2000); Merloni and Piccoli (2001); Landi et al. (2002); Bondi et al. (2002); Prosser and Sarzo (2003); Lastrucci et al. (2004, 2005, 2007, 2010, b, 2012); Maiorca et al. (2007); Tasinazzo (2009); Assini et al. (2010)
	paspaletosum paspaloidis	–	X	–	Biondi et al. (2002, 2013)
40	*Caricetum pendulo-panormitanae*	–	–	X	Conti et al. (2005); Gianguzzi et al. (2013)
41	*Phytocoenon* with *Veronica anagallis-aquatica*	–	X	–	Gansi (2007–2008); Lasen (2014) *in verbis*
	Caricion gracilis		–		
42	*Caricetum gracilis* [incl. phytocoenon with *C. gracilis*, Groupment with *Carex riparia* and *C. gracilis*]	X	–	–	Pignatti (1953); Pedrotti and Sanesi (1969); Cortini Pedrotti et al. (1973a); Pedrotti (1979, 1980, 1988b, 2004a, 2008); Lausi and Gérdol (1980); Bracco (1981); Pedrotti and Cortini Pedrotti (1982); Venanzoni and Canullo (1984); Marchiori and Sburlino (1986); Pedrotti and Venanzoni (1987); Pirone (1987) (1997); Canullo et al. (1988); Venanzoni (1988a, b, 1995); Gèhu and Biondi (1988); Poldini (1989); Buchwald (1994); Buffa et al. (1995); Bracco et al. (1998, 2000); Scortegagna and Tasinazzo (2002); Prosser and Sarzo (2003, 2004); Gigante and Venanzoni (2004); Pellizzari et al. (2005); Gansi (2007–2008); Tasinazzo (2009); Cianfaglione (2009); Cianfaglione (2012)
	typicum	X	–	–	Cortini Pedrotti et al. (1973a)
	caricetosum buxbaumi	X	–	–	Cortini Pedrotti et al. (1973a)
43	*Caricetum buxbaumii*	X	–	–	Pedrotti and Sanesi (1969); Cortini Pedrotti et al. (1973a); Pedrotti (1982); Buchwald (1994); Venanzoni and Gigante (2007); Pirone and Ciaschetti, [not published releves]
44	*Ranunculo apennini-Caricetum buxbaumii*	X	–	–	Pedrotti (2015, 2016)

(continued)

Table 8.1 (continued)

	Communities	AC	CM	M	Literature
45	*Caricetum paniculatae*	X	–	–	Pedrotti (1963, 1965, 1969, 1974, 1983, 1995, 2000); Pirone (1987); Venanzoni (1988b); Balátová-Tuláčková and Venanzoni (1990); Boiti et al. (1990); Corbetta and Pirone (1989); Buchwald (1994); Tammaro (1995); Frattaroli (1994); Pirone (1997); Pirone (1997); Corbetta et al. (1998); Pirone and Corbetta (1998); Biondi et al. (1999); Venanzoni and Gigante (2000); Minghetti and Pedrotti (2000); Luchetta (2001); Andrich et al. (2001); Venanzoni et al. (2003); Presti et al. (2004); Vanacore Falco and Venanzoni (2007–2008, 2009); Venanzoni (2014)
46	*Caricetum ripariae* [Incl. *Carex riparia*, community, Groupment with *Carex riparia* and *Carex gracilis*; *Galio palustris-Caricetum ripariae* Balátová-Tuláčková in Balátová-Tuláčková Mucina Ellemauer et Wallnöfer in Grabherr et Mucina 1993]	–	X	–	Pignatti (1953); Pirola (1968); Gerdol et al. (1979); Barbagallo et al. (1979a, b); Pedrotti (1980); Sartori (1981); Avena and Scoppola (1987); Gèhu and Biondi (1988); Balátová-Tuláčková and Venanzoni (1989); Minissale and Spampinato (1995); Sburlino et al. (1995); Brullo et al. (1996, 2002); Pirone et al. (1997); Corbetta et al. (1998); Pirone and Corbetta (1998); Piccoli (1998); Sarzo et al. (1999); Sarzo et al. (1999); Venanzoni and Gigante (2000, 2004); Tomaselli (2001); Pellizzari and Piccoli (2001); Venanzoni et al. (2003); Prosser and Sarzo (2004); Gigante and Venanzoni (2004); Pellizzari et al. (2004, 2005); Tomaselli et al. (2006); Argenti et al. (2006); Gansi (2007–2008); Bolpagni et al. (2007); Lastrucci et al. (2008, 2010a, b); Cianfaglione (2009); Lasen (2014) *in verbis*; Sburlino (2014) *in verbis*; Venanzoni (2014)
47	*Caricetum vesicariae*	X	–	–	Cortini Pedrotti et al. (1973a); Aita et al. (1976); Montanari and Guido (1980); Andreis and Rodondi (1982); Badino et al. (1982); Gérdol (1987); Pirone (1987, 1997); Venanzoni (1987); Venanzoni (1988b, 1995); Rey (1990); Buchwald (1994); Bracco and Nola (1995); Gerdol and Tomaselli (1997); Luchetta (2001); Prosser and Sarzo (2003); Gigante and Venanzoni (2004); Tomaselli et al. (2006); Bolpagni et al. (2007); Pedrotti and Tisi et al. (2007); Gargano et al. (2007); Lastrucci et al. (2008); Pedrotti e Venanzoni, dati inediti
48	*Caricetum vulpinae*	X	–	–	Cortini Pedrotti et al. (1973a); Badino et al. (1982); Pedrotti (1982); Pirone (1987, 1997); Caniglia and Basso (1995)

49	*Caricetum distichae*	X	–	–	Pedrotti and Venanzoni (1987); Venanzoni (1993); Pedrotti (2016); Venanzoni [not published releves]
50	*Phalaridetum arundinaceae*	–	X	–	Pedrotti and Sanesi (1969); Bracco (1981); Buchwald (1984); Bracco et al. (1984); Marchiori and Sburlino (1986); Pedrotti and Venanzoni (1987); Pedrotti (1988); Poldini (1989); Andreis et al. (1993); Sburlino et al. (1995); Pirone e Corbetta (1998); Venanzoni and Gigante (2000, 2004); Prosser and Sarzo (2003); Cristea et al. (2004); Gigante e Venanzoni (2004); Pellizzari et al. (2005); Brusa et al. (2006b); Gansi (2007–2008); Lastrucci et al. (2007, 2010a, b, 2012); Ceschin and Salerno (2008); Venanzoni (2014) *in verbis*
51	*Carici hispidae-Juncetum subnodulosi*	–	–	X	Gèhu and Biondi (1988)
52	Groupement with *Carex melanostachya*	–	X	–	Tasinazzo (2009)
	***Carici-Rumicion hydrolapathi* Passarge**				
53	*Cicuto virosae-Caricetum pseudocyperi*	X	–	–	Pedrotti (2004); Badino et al. (1982); Anoè and Caniglia (1987); Tomaselli et al. (2006)
	caricetosum paniculatae	X	–	–	Tomaselli et al. (2006)
54	*Mentho aquaticae-Caricetum pseudocyperi*	–	X	–	Orsomando and Pedrotti (1986); Venanzoni and Gigante (2000, 2004); Corbetta and Pirone (1989); Ceschin and Salerno (2008)
	Hyperico hircini--Caricion microcarpae	]			
55	*Hyperico hircini-Caricetum microcarpae*	–	–	X	Bacchetta and Mossa (2002, 2004); Bacchetta et al. (2007), Mossa et al. (2008)
	oenanthetosum crocatae	–	–	X	Bacchetta and Mossa (2004)
	adiantetosum capilli-veneris	–	–	X	Bacchetta and Mossa (2004)

Annex 1

The *Caricion gracilis* alliance includes tall sedge-communities growing on waterlogged mineral to more or less organic soil. Its distribution includes inland Italian territories as Alps, Po Valley and Apennine karstic highlands where most of the known European associations are present and where vegetation reaches the maximum of biodiversity due to the relict presence of rare artic-alpines species as *Carex buxbaumi*, *C. disticha* (Pedrotti 1971, 1977, 1985) and *C. vulpina* (Costalonga and Venanzoni 2009). A modest floristic differentiation is given by the presence of polymorphics species such as those belonging to *Ranunuculus auricomus* aggregate (Masci et al. 1994; Dunkel 2011) and *Taraxacum palustre* aggregate (Aquaro *et al.* 2008; Peruzzi and Carlesi 2012). From syntaxonomical point of view no others significant differences are observed comparing to the central European classification.

The *Carici-Rumicion hydrolapathi* alliance includes tall sedges and tall-herb vegetation that grows on coarse and fine organic sediments originating often from stolons, rhizomes and adventitious roots of the same species. The colonization of these dense systems of roots and organic sediments by communities of this alliance can form floating mats contributing to the process of terrestrialization. This alliance has mainly central European distribution with occurrences in the Italian peninsula from the Alps to large natural lakes in the Center and South of the peninsula (e.g. lakes of Chiusi, Trasimeno, Piediluco and Monticchio) (Venanzoni and Gigante 2000). In this ecological situation it is also possible to found communities dominated by *Phragmites australis* and *Cladium maricus* capable of growing on their own floating rhizomes. From syntaxonomical point of view there are moderate differences with Central European classification due to the presence of one endemic association.

Caricion microcarpae Gamisans (1968) 1975 alliance is characterized by *Carex microcarpa* endemic species of Sardinia, Corsica and Tyrrenian part of Central Italy (Conti et al. 2005). It was described in Corsica by Gamisans in 1968 and re-used in 1975 by the same Author to include communities of tall-herbs with *Gentiana asclepiadea* and woods with *Alnus glutinosa* and *A. cordata*. This alliance described with a doubtful attribution to higher units of *Populetalia* and not used in similar contexts (Foggi et al. 2006; Angius and Bacchetta 2009), is probably an abandoned syntaxon (Bardat et al. 2001, 2004). Bacchetta and Mossa (2004) and Mossa et al. (2008) took up this alliance to include communities of *Carex microcarpa* attributable to marsh tall-sedge vegetation but in a different syntaxonomical (compared to the original description) context inside the *Magnocaricetalia* Order. This alliance should be discussed taking into account the new distribution data of the endemic species *Carex panormitana* in Sicily and Sardinia (Urbani et al. 2013) and formalized according to the nomenclatural code (Weber et al. 2000).

In conclusion first it can be observed that most of the associations of the Order *Magnocaricetalia* described in Europe are present in Italian Regions (Table 8.2) and a low presence of endemic syntaxa is observed. Second a formalization according the Phytosociological code is still necessary.

Table 8.2 Regional distribution of syntaxa (x = presence, ? = doubtful presence, y = presence deduced from floristic data in Conti et al. 2005)

	Regional distribution of investigated Plant Communities	VDA	PIE	LOM	TAA	VEN	FVG	LIG	EMR	TOS	MAR	UMB	LAZ	ABR	MOL	CAM	PUG	BAS	CAL	SIC	SAR
1	*Caricetum acutiformis*	y	x	x	x	x	–	–	x	x	–	x	x	x	?	–	–	–	–	–	–
2	*Caricetum appropinquatae*	x	x	x	x	–	–	–	–	–	–	–	–	–	–	–	–	–	–	–	–
3	*Caricetum buxbaumii*	–	y	–	y	–	y	–	–	–	–	x	–	x	–	–	–	–	–	–	–
4	*Caricetum caespitosae*	?	y	y	?	–	y	x	–	–	–	–	–	–	–	–	–	–	–	–	–
5	*Caricetum diandrae*	x	–	–	x	x	–	–	–	–	–	–	–	–	–	–	–	–	–	–	–
6	*Caricetum distichae*	–	–	–	x	–	–	–	–	–	–	x	–	x	–	–	–	–	–	–	–
7	*Caricetum elatae*	x	x	x	x	x	x	x	x	x	x	x	x	x	x	y	–	–	y	–	–
8	*Caricetum gracilis*	–		X	x	x	x		x		x	x		x	x		x		x	–	–
9	*Caricetum hispidae*	–	–	–	–	–	–	–	–	–	–	–	–	–	–	–	x	–	–	x	–
10	*Caricetum otrubae*	–	–	–	–	–	–	–		x	–	–	–	x	–	–	–	–	–	x	–
11	*Caricetum paniculatae*	x	y	x	x	x	y	y	x	y	x	x	x	x	x	–	–	x	x	y	–
12	*Caricetum pendule-panormitanae*	–	–	–	–	–	–	–	–	–	–	–	–	–	–	–	–	–	–	x	y
13	*Caricetum ripariae*	–	x	x	x	y	–	–	x	x	x	x	x	x	?	y	x	x	–	x	–
14	*Caricetum vesicariae*	x	x	x	x	x	–	x	x	x	x	x	x	x	–	–	–	–	x	–	–
15	*Caricetum vulpinae*	?	–	–	–	–	y	–	–	–	–	x	–	x	–	–	–	–	–	–	–
16	*Carici distantis-Schoenetum nigricantis*	–	–	–	–	–	–	–	–	–	–	–	–	–	–	–	–	–	–	x	–
17	*Carici elatae-Calamagrostietum canescentis*	–	x	x	x	–	–	–	–	–	–	–	–	–	–	–	–	–	–	–	–
18	*Cicuto virosae-Caricetum pseudocyperi*	–	x	x	x	x	–	–	–	–	–	–	–	–	–	–	–	–	–	–	–
19	*Cladietum marisci*	x	x	x	x	x	x		x	x	x	x	x	x	x	y	x	x	–	–	x
20	*Cyperetum longi*	–	–	x	–	–	–	–	–	x	–	x	x	x	–	–	–	x	–	x	–
21	*Cypero-Caricetum cuprinae*	–	–	–	–	–	–	–	–	–	y	x	–	–	–	x	–	–	–	x	–
22	*Eleocharitetum palustris*	x	–	x	x	x	x		x	x	x	x	–	–	–	–	–	–	x	–	–
23	*Eleocharitetum uniglumis*	y	–	–	y	–	–	x	x	–	–	–	–	–	–	–	–	–	–	–	–
24	*Equiseto palustris-Juncetum effusi*	–	–	–	–	–	–	–	–	–	–		x	–	–	–	–	–	–	x	–
25	*Eupatorio cannabini-Caricetum elatae*	–	–	–	–	–	–	–	–	–	y	x	–	–	–	–	–	–	–	–	–
26	*Galio palustris-Caricetum rostratae*	x	x	x	x	x	x	x	x	x	–	–	?	x	–	–	–	–	x	–	–
27	*Galio palustris-Juncetum inflexi*	–	–	–	–	–	–	–	–	–	x	x	x	x	–	–	–	–	–	x	–
28	Groupment with *Carex melanostachya*	–	–	–	–	x	–	–	–	–	–	–	–	–	–	–	–	–	–	–	–
29	Groupement with *Juncus effusus* and *Galium elongatum*	–	–	–	–	–	–	–	–	x	–	–	–	–	–	–	–	–	–	–	–
30	Groupement with *Juncus subnodulosus*	–	–	–	–	x	–	–	–	x	–	x	x	–	–	–	–	–	–	–	–

(continued)

Table 8.2 (continued)

	Regional distribution of investigated Plant Communities	VDA	PIE	LOM	TAA	VEN	FVG	LIG	EMR	TOS	MAR	UMB	LAZ	ABR	MOL	CAM	PUG	BAS	CAL	SIC	SAR
31	Groupement with *Menyanthes trifoliata*	–	–	x	x	–	–	–	x	–	–	–	–	–	–	–	–	–	x	–	–
32	Groupment with *Lysimachia vulgaris*	–	–	–	–	x	–	–	–	–	x	–	–	–	–	–	–	–	–	–	–
33	*Holoschoeno-Juncetum subnodulosi*	–	–	–	–	–	–	–	x	–	–	–	–	–	–	–	x	–	–	–	–
34	*Hyperico hircini-Caricetum microcarpae*	–	–	–	–	–	–	–	–	–	–	–	–	–	–	–	–	–	–	–	x
35	*Hyperico tetrapteri - Caricetum acutiformis*	–	–	–	–	–	–	–	–	–	–	–	–	x	–	–	–	–	–	–	–
36	*Iridetum pseudacori*	–	–	–	–	x	–	–	x	x	–	–	x	–	x	–	x	–	–	–	–
37	*Junco maritimi-Cladietum marisci*	–	–	–	–	–	–	–	–	–	–	–	–	–	–	–	x	–	–	–	–
38	*Leersio oryzoidis-Juncetum effusi*	–	–	–	–	–	x	–	–	x	–	x	x		–	–	–	–	–	–	–
39	*Leucojo-Caricetum elatae*	–	–	x	–	–	–	–	x	–	–	–	–	–	–	–	–	–	–	–	–
40	*Leucojo-Caricetum ripariae*	–	–	x	–	–	–	–	x	–	–	–	–	–	–	–	–	–	–	–	–
41	*Mentha aquatica* Community	–	–	–	–	–	–	x	–	x	–	–	–	–	–	–	–	–	–	–	–
42	*Mentho aquaticae-Caricetum pseudocyperi*	–	–	–	–	–	–	–	–	x	–	x	x	x	–	–	–	–	–	–	–
43	*Mentho aquaticae-Typhetum minimae*	–	–	–	–	–	–	–	–	–	–	x	–	–	–	–	–	–	–	–	–
44	*Peucedano palustris-Caricetum lasiocarpae*	y	y	x	x	y	y	–	–	–	–	–	–	–	–	–	–	–	–	–	–
45	*Phalaridetum arundinaceae*	–	x	x	x	x	x	–	–	x	x	x	x	–	–	–	–	–	–	–	–
46	*Carici hispidae-Juncetum subnodulosi*	–	–	–	–	–	–	–	–	–	–	–	–	–	–	–	x	–	–	–	–
47	*Ranunculo apenninae-Caricetum buxbaumii*	–	–	–	–	–	–	–	–	–	–	x	–	–	–	–	–	–	–	–	–
48	*Ranunculus flammula – Juncus articulatus* Community	–	–	–	–	–	–	–	–	–	–	–	–	–	–	–	–	–	x	–	–
49	*Soncho maritimi-Cladietum marisci*	–	–	–	–	x	x	–	x	–	x	–	x	–	–	–	x	–	–	x	x
50	*Teucrio scordii-Juncetum subnodulosi*	–	–	–	–	–	–	–	x	–	–	–	–	–	–	–	–	–	–	–	–
51	*Thelypterido palustris-Phragmitetum australis*	–	–	x	x	–	–	–	–	x	–	–	x	–	–	–	–	–	–	–	–
52	*Veronica anagallis-aquatica* phytocoenon	–	–	x	–	–	–	–	–	–	–	–	–	–	–	–	–	–	–	–	–
53	*Filipendulo ulmariae-Caricetum randalpinae*	–	–	–	–	x	–	–	–	–	–	–	–	–	–	–	–	–	–	–	–
54	Groupment with *Equisetum fluviatile*	–	–	–	–	–	–	x	–	–	–	–	–	–	–	–	–	–	–	–	–
55	Groupment with *Veronica beccabunga*	x	–	–	–	–	–	–	–	–	–	–	–	–	–	–	–	–	–	–	–

The *Magnocaricetalia* vegetation by occupying a transitional position, compared to *Phragnitetalia* ones, between water and inland territories, takes an important role both from the ecological services and biodiversity conservation point of view. Nonetheless such territories are strongly threatened because of the high anthropic pressure as alteration of lake shores and river banks, eutrophication processes, agriculture and urban expansion. Despite their biodiversity and ecological role, wet areas are only scarcely protected from the vegetation point of view although there are many directives about conservation of water bodies and wetlands (ISPRA 2011). Annex I of the Habitats 92/43/EEC according with the on-line interpretation manual (Biondi et al. 2012) offers tools for protection only of *Cladietum marisci* association but the risk of fragmentation and estinction is hight for all communities of the Order (Pedrotti 1983, 2013) and considering the hight level of vulnerability they should be included almost all in a Red List of Italian vegetation.

Annex 1

Literature Cited in Table 8.1

Aita L, Martini E, Orsino F (1976) Flora e vegetazione dei laghetti delle Agoraie e zone limitrofe (Appennino Ligure orientale) Lav Soc Biogeogr 6:163–229

Allegrezza M, Mentoni M, Tesei G (2010) Geomorfologia e paesaggio vegetale: l'esempio della grande frana di Pescacci (Comune di Serra San Quirico - Appennino centrale). Fitosociologia 47(2):57–97

Andreis C, Cerabolini B (1995) La Brughiera Briantea: la vegetazione e il Piano di Gestione. Coll Phytosoc 21:195–224

Andreis C, Lazzaroni L, Rodondi G, Zavagno F (1993) La vegetazione delle torbiere del Sebino e le direttive del piano di gestione. Coll Phytosoc XXI:511–546

Andreis C, Rodondi G (1982) La torbiera di Pian di Gembro (Prov. Di Sondrio). Collana Programma Finalizzato "Promozione della Qualità dell'Ambiente". CNR AQ/1/221

Andreis C, Zavagno F (1996) La vegetazione del Lago di Ganna, con particolare riferimento ai rapporti spaziali tra le cenosi dei *Molinietalia* e degli *Scheuchzerietalia palustris*. Il Naturalista Valtellinese. Atti Mus Civ Stor Nat Morbegno 7:33–56

Andreucci F, Castelli M, [2006 (2002)] Alcuni aspetti di vegetazione erbacea nelle aree golenali del torrente Scrivia (Piemonte, Italia settentrionale). Arch Geobot 8(1–2):49–68

Angiolini C, Casini F, Centi C, Frignani F, Landi M (2006) Valutazione degli elementi floristico-vegetazionali, habitat e paesaggio colturale per la realizzazione del piano strutturale del Comune di Monteroni d'Arbia. Università degli Studi di Siena Dipartimento di Scienze Ambientali "G. Sarfatti" Sezione di Ecologia e Sistematica Animale e Vegetale

Angiolini C, Chiarucci A, De Dominicis V, Gabellini A, Morrocchi D, Selvi F (1999) Lineamenti vegetazionali dell'Area Naturale Protetta del Fiume Elsa. Atti Accad Fisiocritici Siena 18:101–122

Angiolini C, Landi M, Boddi M, Frignani F (2005) La vegetazione dell'alveo fl uviale del Sito di Importanza regionale Torrente Trasubbie (Grosseto, Toscana Meridionale). Atti Soc Tosc Nat Mem Ser B 122:127–151

List of Authors Cited in Table 8.1

Anoè N, Caniglia G (1987) La vegetazione acquatica e palustre di alcune cave di argilla dell'entroterra veneziano. Lav Soc Ven Sci Nat Venezia 12:159–175

Anzalone B, Iberite M, Lattanzi E (2010) La Flora vascolare del Lazio. Inf Bot Ital 42(1):187–317

Argenti C, da Pozzo M, Lasen C (2006) Segnalazioni Floristiche per la Provincia di Belluno III. Ann Mus Civ Rovereto Archeol Stor Sci Nat 21:167–206

Arrigoni PV, Riccieri C (1981) La vegetazione dei laghi di Chiusi e di Montepuliano (Siena). Atti Soc Tosc Sci Nat Mem Serie B 88:285–299, tabb. 10

Arrigoni PV (1990) Flora e vegetazione della Macchia lucchese di Viareggio (Toscana). Webbia 44(1):1–62

Arrigoni PV, Nardi E, Raffaelli M (1985) La vegetazione del parco naturale della Maremma (Toscana). Università degli studi di Firenze, Dip. di Biologia Vegetale

Arrigoni PV, Papini P (2003) La vegetazione del sistema fluviale Lima - Serchio (Toscana Settentrionale). Parlatorea 6:95–129

Assini S, Bracco F, Carrea A, Villani M (2010) Stato delle conoscenze floristico-vegetazionali sul corso planiziale del Fiume Po. Bio Ambient 24(1):1–18. ISSN: 1129-504X

Avena GC, Scoppola A (1987) Valutazione della situazione ambientale del Lago di Nemi. Caratteristiche dei complessi ad idrofite ed elofite. Univ. "La Sapienza", Assessorato alla Sanità e all'Ambiente, Roma, pp 75–94

Bacchetta G, Casti M, Zavattero L (2007) Analisi della vegetazione del distretto minerario di Montevecchio (Sardegna sud-occidentale). Fitosociologia 44 (2):83–108

Bacchetta G, Mossa L (2004) Studio fitosociologico delle cenosi a Carex microcarpa Bertol. ex Moris della Sardegna meridionale. Fitosociologia 41(1) suppl. 1:171–178

Badino G, Camoletto R, Dal Vesco G (1982–83) Popolamenti fanerogamici del bacino di Candia e assetto idrobiologico del lago. Rev Valdòtaine d'Hist Nat. 36–37:43

Balátová-Tuláčková E (1963) Zur Systematik der europäischen Phragmitetea. Preslia 35:118–122

Balátová-Tuláčková E (1991) Das Cladietum marisci. Veröff Geobot Inst ETH 106:7–34

Balátová-Tuláčková E (1994) Magnocaricion elatae-Gesellschaften - Eine Ergänzung zum Werk "Die Pfl anzengesellschaften Österreichs". Verh Zool Bot Ges Österreich 131:27–36

Balátová-Tuláčková E, Mucina L, Ellmauer T, Wallnöfer S (1993) Phragmiti-Magnocaricetea. In: Grabherr G, Mucina L (eds) Die Pfl anzengesellschaften Österreichs, Teil 2, Natürliche waldfreie Vegetation. Gustav Fischer Verlag, Jena, pp 79–130

Balátová-Tuláčková E, Pavlíček V (1980) Beitrag zur Kenntnis der chemischen Zusammensetzung der Phragmiti-Magnocaricetea und der Agropyro-Rumicion Arten. Folia Geobot Phytotax 15:225–244

Balátová-Tuláčková E, Venanzoni R (1989) Sumpf und Feuchtrasengesellschaften in der Verlandungszone des Kalterer Sees (Lago di Caldaro), der Montiggler (Monticolo) Seen und in der Etsch (Adige) Aue, Oberitalien. Folia Geobot Phytotax 24:253–295

Balátová-Tuláčková E, Venanzoni R (1990) Beitrag zur Kenntniss der Nass- und Feuchtwiesen in der montanen Stufe der Provinz Bozen (Bolzano), Italien. Tuexenia 10:153–171

Baldoni M, Biondi E (1993) La vegetazione del medio e basso corso del Fiume Esino (Marche - Italia centrale). Stud Bot 11:209–257

Ballelli S, Biondi E, Cortini Pedrotti C, Francalancia C, Orsomando E, Pedrotti F (1981) Il patrimonio vegetale delle Marche. Regione Marche, Assessorato Urbanistica Ambiente, Ancona, pp 182–189

Barbagallo C, Brullo S, Fagotto F (1979a) Vegetazione a Platanus orientalis L. e altri aspetti igrofili dei fiumi Iblei (Sicilia meridionale). Pubblicazioni dell'Istituto di Botanica dell'Universita di Catania

Barbagallo C, Brullo S, Furnari F (1979b) Osservazioni fitosologiche sulla vegetazione del fiume Ciane (Sicilia orientale). Pubblicazioni dell'Istituto di Botanica dell'Universita di Catania

Barbagallo C, Brullo S, Furnari F (1979c) Su alcuni aspetti di vegetazione igrofila di Serra del Re (Monti Nebrodi). Pubblicazioni dell'Istituto di Botanica dell'Universita di Catania

Barberis G, Mariotti M (1981) Ricerche geobotaniche sulle zone umide del Gruppo di Voltri (Appennino Ligure-Piemontese). Estratto da: Archivio Botanico Biogeografico ltaliano, n. 57 ½

Bartolo G, Brullo S, Marcerò C (1982) La vegetazione costiera della Sicilia sud-orientale. Quaderni C.N.R. serie AQ/1/226: 1–49

Bertolani Marchetti D, Accorsi CA, Arobba D, Bandini Mazzanti M, Bertolani M, Bertolani R, Boni L, Braggio G, Cellai Ciuffi G, De Cunzo T, Ferrari C, Forlani L, Guido M, Paoli P, Rodolfi G (1977) Studi ecologici e paleoecologici nella palude della Chioggiola presso Pavullo nel Frignano. Estratto dal vol. "Pavullo e il Medio Frignano". Modena, Aedes Muratoriana (Biblioteca Nuova Serie N. 38)

Biondi E (1986) La vegetazione del Monte Conero (con carta della vegetazione alla scala 1:10.000). Regione Marche, Ancona, pp 1–95

Biondi E, Baldoni M (1994) La vegetazione del Fiume Marecchia (Italia centrale). Biogeographia 17:51–87

Biondi E, Baldoni M, Formica E (1995) Memoria illustrativa sugli habitats "Lagoons" e "Salt steppes". In: Habitat prioritari. Società Botanica Italiana. http://vnr.unipg.it/habitat/

Biondi E, Blasi C (2014) Prodromo della Vegetazione Italiana. Check-list sintassonomica aggiornata di classi, ordini e alleanze presenti in Italia. Società Botanica Italiana. Available online at http://www.prodromo-vegetazione-italia.org. Accessed 30 Apr 2014

Biondi E, Bracco F, Nola P (eds) (1996) Bibliografia fitosociologica italiana (1909–1993). Fitosociologia 31:5–10

Biondi E, Bracco F, Nola P (eds) (1997b) Lista delle unità sintassonomiche della vegetazione italiana. Fitosociologia 33:5–11

Biondi E, Calandra R, Gigante D, Pignattelli S, Rampiconi E, Venanzoni R (2002) Il paesaggio vegetale della provincia di Terni. Provincia di Terni. Università di Perugia.

Biondi E, Casavecchia S, Guerra V (2006) Analysis of vegetation diversity in relation to the geomorfological characteristics in the Salento coasts (Apulia, Italy). Fitosociologia 43(1):25–38

Biondi E, Casavecchia S, Radetic, Z (2002) La vegetazione dei "guazzi" e il paesaggio vegetale della pianura alluvionale del tratto terminale del Fiume Musone (Italia centrale). Fitosociologia 39(1):45–70

Biondi E, Catorci A, Pandolfi M, Casavecchia S, Pesaresi S, Galassi S, Pinzi M, Vitanzi A, Angelini E, Bianchelli M, Cesaretti S, Foglia M, Gatti R, Morelli F, Paradisi L, Ventrone F, Zabaglia C (2007) Il Progetto di "Rete Ecologica della Regione Marche" (REM): per il monitoraggio e la gestione dei siti Natura 2000 e l'organizzazione in rete delle aree di maggiore naturalità. Fitosociologia 44(2) Suppl. 1:89–93

Biondi E, Giustini A, Taffetani F. (1994) La vegetazione di Portonovo (Ancona): analisi, rilevamento cartografico e proposte di recupero. Verde, Città e Territorio, Ancona, pp 197–208

Biondi E, Vagge I, Baldoni M, Taffetani F (1997a) La vegetazione del Parco Fluviale Regionale del Taro (Emilia Romagna). Fitosociologia 34:69–110

Biondi E, Vagge I, Baldoni M, Taffetani F (1999) La vegetazione del Parco Fluviale Regionale dello Stirone (Emilia Romagna). Fitosociologia 36:67–93

Biondi E, Vagge I, Baldoni M, Taffetani F (2004) Biodiversità fitocenotica e paesaggistica dei fiumi dell'Italia centro-settentrionale: aspetti fitosociologici e sinfitosociologici. Studi Trent Sci Nat Acta Biol 80:13–21

Bonomi C, Buffa G (2000) (1997) Valutazione della qualità della vegetazione della torbiera 21 delle Viote del Monte Bondone (TN) con lineamenti di corretta gestione ambientale. Studi Trent Sci Nat Acta Biol 74:77–97

Bracco F (1992) Variazioni del paesaggio vegetale nella golena del fiume Po. Colloques phytosociologiques Ecologia del paesaggio, Camerino

Bracco F (1981) Note sulla vegetazione acquatica e palustre della bassa valle del Ticino. Not Fitosoc 17:55–68

Bracco F (2001) Il primo aggiornamento della Lista sintassonomica italiana. Fitosociologia Suppl. 1, 38(2):5–11

Bracco F, Buffa G, Ghirelli L, Sburlino G, Zuccarello V (1998) The phytosociological information and the management of the upspring vegetation of river Sile Regional Park (Venetian plain - Northern Italy). Arch Geobot 4:51–57

Bracco F, Buffa G, Ghirelli L, Sburlino G, Zuccarello V (2000) The phytosociological information and the management of the upspring vegetation of the River Sile regional Park (Venetian plain – Northern Italy). Arch Geobot 4:51–57

Bracco F, Nola P (1995a) Project LISY: data banks for the compilation of a national list of syntaxonomic units of Italy. Ann Bot (Roma) 53:163–169

Bracco F, Nola P (1995b) La vegetazione acquatica e palustre dei lagoni di Mercurago. Fitosociologia 29:137–150

Bracco F, Panfili E, Sburlino G, Venanzoni R (2007) Il secondo aggiornamento della Lista sintassonomica italiana (2000–2004). Fitosociologia Suppl. 1, 44(1): 5–12.

Bracco F, Sartori F, Terzo V (1984) Indagine geobotanica per la valutazione di un'area della bassa padania occidentale. Atti Ist Bot Lab Critt 3(serie 7):5–50

Brullo S (1978) La vegetazione palustre di Capo Feto. Un ambiente da salvare. Ambienti umidi costieri. Atti II Conv. Siciliano di ecologia, Noto, pp 23–25

Brullo S, De Santis C, Furnari F, Longhitano N, Ronsisvalle GA (1988) La vegetazione dell'oasi della foce del Simeto (Sicilia orientale). Braun-Blanquetia 2:165–188

Brullo S, Furnari F (1976) Le associazioni vegetali degli ambienti palustri costieri della Sicilia. Not Fitosoc 11:1–43

Brullo S, Furnari F, Scelsi F (1993) Considerazioni fitosociologiche sulla vegetazione di Cava d'Ispica (Sicilia meridionale). Boll Acc Gioenia Sci Nat Catania 26(341):49–83

Brullo S, Giusso Del Galdo G, Minissale P, Siracusa G, Spampinato G (2002) Considerazioni sintassonomiche e fitogeografiche sulla vegetazione della Sicilia. Boll Acc Gioenia Sci Nat Catania 35(361):325–359

Brullo S, Grillo M (1978) Ricerche fitosociologiche sui pascoli dei monti Nebrodi (Sicilia settentrionale). Not Fitosoc 13:23–61

Brullo S, Grillo M, Guglielmo A (1996) Considerazioni fitogeografiche sulla flora iblea. Boll Acc Gioenia Sci Nat 29:45–111

Brullo S, Grillo M, Scalia C (1980) Su una nuova associazione dei Pegano-Salsoletea osservata presso Porto Empedocle (Agrigento). Giorn Bot Ital 114(1–2):43–52

Brullo S, Minissale P, Siracusa G (1998a) Quadro sintassonomico della vegetazione iblea. Boll Acc Gioenia Sci Nat 29(352):113–150

Brullo S, Minissale P, Spampinato G (1994) Studio fitosociologico della vegetazione lacustre dei Monti Nebrodi (Sicilia settentrionale). Fitosociologia 27:5–50

Brullo S, Ronsisvalle GA (1975) La vegetazione dei Gorghi Tondi e del lago Preola, presso Mazara del Vallo (Sicilia occidentale). Not Fitosoc 10:45–67

Brullo S, Scelsi F, Spampinato G (1998b) Considerazioni sintassonomiche sulla vegetazione perenne pioniera dei substrati incoerenti dell'Italia meridionale e Sicilia. Itinera Geobot 11:403–424

Brullo S, Sciandrello S (2006) La vegetazione del bacino lacustre "Biviere di Gela" (Sicilia meridionale). Fitosociologia 43(2):21–40

Brullo S, Siracusa G (1996) Studio fitosociologico dell'isola di Linosa. Doc Phytosoc 16:123–164

Brullo S, Siracusa G (2000) Indagine fitosociologica su di un'area umida del versante sud-occidentale dell'Etna di notevole interesse naturalistico. Arch Geobot 4(1):71–90

Brullo S, Spampinato G (1990) La vegetazione dei corsi d'acqua della Sicilia. Boll Acc Gioenia Sci Nat 23:119–252

Brusa G, Castiglioni L, Cerabolini B (2006a) La vegetazione dell'istituenda Riserva naturale Oasi di Lacchiarella (Parco agricolo sud Milano). Pianura. Scienze e storia dell'ambiente padano n. 20:5–41

Brusa G, Raimondi B, Cerabolini B (2006b) La vegetazione della Riserva Naturale "Lago di Biandronno" (Lombardia, Italia Settentrionale). Fitosociologia 43(2):111–128

Buchwald R (1992) Il Veronico-Apietum submersi, una nuova associazione dell'Italia centrale. Doc Phytosoc 14:513–529

Buchwald R (1994) Vegetazione e odontofauna negli ambienti acquatici dell'Italia centrale. Braun-Blanquetia 11:1–77

Buffa G, Ghirelli L, Sburlino G (1995) La vegetazione delle sorgenti del fiume Sile (Veneto - Italia nord-orientale). Giorn Bot Ital 129:265

Caniglia G, Basso F (1995) Descrizione del paesaggio vegetale del corso del fiume Merzenego (pianura veneta) mediante un approccio sinfitosociologico. Quad Staz Ecol Civ Mus St Nat Ferrara 9:137–153

Caniglia G, Mondin F, Carpenè B (1992) Aspetti floristico-vegetazionali di un lobo di meandro del Parco del Sile (S. Michele Vecchio, Venezia). Lav Soc Ven Sci Nat Venezia 17:151–173

Canullo R, Pedrotti F, Venanzoni R (1988) I prati umidi ed inondati dell'alto Trigno (Molise, Italia). Doc Phytosoc 11:583–606

Canullo R, Pedrotti F, Venanzoni R (1991) Les processus dynamiques dans la végétation de la tourbière de Fiavé (Italie du nord). Phytocoenosis, 3(N.S.) Suppl Cartogr Geobot 2:189–194

Canullo R, Pedrotti F, Venanzoni R, (1994) La Torbiera di Fiavé. Guida all'escursione della Società Italiana di Fitosociologia in Trentino

Canullo R, Venanzoni R (1988) Studio preliminare della struttura di una popolazione di *Cytisus sessilifolius* L. nella Riserva naturale di Torricchio (Appennino centrale, Italia). S.IT.E. Atti del terzo congresso nazionale della Società Italiana di Ecologia 7:761–765

Catorci A, Orsomando E, Silvi B (1996) Il paesaggio vegetale del Foglio Nocera Umbra. N. 312 della Carta d'Italia I.G.M., in Scala 1: 50.000. Coll Phytosoc 24:665–673

Ceschin S, Salerno G (2008) La vegetazione del basso corso del Fiume Tevere e dei suoi affl uenti (Lazio, Italia). Fitosociologia 45:39–74

Cianfaglione K (2009) The hygrophilous vegetation of the Sulmona basin (Abruzzo, Italy). Contrib Bot 44:49–56

Conti F, Abbate G, Alessandrini A, Blasi C (eds) (2005) An annotated check-list of the Italian vascular flora. Palombi ed., Roma

Conti F, Alessandrini A, Bacchetta G, Banfi E, Barberis G, Bartolucci F, Bernardo L, Bonacquisti S, Bouvet D, Bovio M, Brusa G, Del Guacchio E, Foggi B, Frattini S, Galasso G, Gallo L, Gangale C, Gottschlich G, Grünanger P, Gubellini L, Iiriti G, Lucarini D, Marchetti D, Moraldo B, Peruzzi L, Poldini L, Prosser F, Raffaelli M, Santangelo A, Scassellati E, Scortegagna S, Selvi F, Soldano A, Tinti D, Ubaldi D, Uzunov D, Vidali M (2007) Integrazioni alla checklist della flora vascolare italiana. Nat Vicentina 10(2006):5–74

Corbetta F, Abbate Frattaroli AR, Pirone G (1998) Vegetazione e specie da conservare S.O.S. verde. Ed. Agricole

Corbetta F, Pirone G (1989) La vegetazione del fiume Tirino (Abruzzo). Arch Bot Ital 65:121–153

Corbetta F, Pirone G (1999) Analisi comparativa della vegetazione delle lagune della costa adriatica e dell'arco jonico pugliese-lucano. Attuale situazione conservazionistica. Boll Mus Civ Stor Nat Venezia (Suppl. 49):135–146

Corsi R (2001) Dinamiche vegetazionali e interventi di gestione nell'ANPIL Bosco di Tanali. In: Venturato E, Petrini R (eds) Lungo le rotte migratorie. Progetti di ricerca sulla vegetazione, l'avifauna e le specie aliene. Quad. Padule Fucecchio n. 1. Centro Ricerca, Documentazione e Promozione Padule Fucecchio

Cortini Pedrotti C, Orsomando E, Pedrotti F, Sanesi G (1973) La vegetazione e i suoli del Pian Grande di Castelluccio di Norcia (Appennino centrale). Atti Ist Bot Lab Critt Univ Pavia IX(serie 6):155–249

Cristea V, Gafta D, Pedrotti F (2004) Fitosociologie. Presa Universitaria Clujana, Cluj-Napoca Currie DJ (1991) Energy and large-sale pattern of animal- and plant- species richness. Am Nat 137(1):27–49

Dell'Uomo A, Agostinelli A (1990) Florula desmidiologica del Trentino - Alto Adige: le torbiere di Nova Ponente e del Doss le Grave. Studi Trent Sci Nat Acta Biol 66(1989):83–111

Ferrari C, Manzi ML (1987) Osservazioni sulla vegetazione macrofitica del Lago Calamone (Appennino settentrionale). Inf Bot Ital 19:82–87

Filigheddu R, Farris E, Biondi E (2000) The vegetation of S'Ena Arrubia lagoon (central-west Sardinia). Fitosociologia 37:39–59

Focarile A (1977) Studio faunistico ed ecologico sulla coleotterofauna di due bacini lacustro-torbosi in Valle d'Aosta. Rev Valdôtaine d'Hist Nat 31:25–54

Foggi B, Gennai M, Gervasoni D, Ferretti G, Rosi C, Viciani D, Venturi E (2007) La carta della vegetazione del SIC Alta valle del Sestaione. Parlatorea IX:41–78

Frattaroli AR, Capranica R (1994) Lineamenti della vegetazione del laghetto e fiume Vetoio (Conca aquilana - Abruzzo, Italia). Micol Veg Mediterr 9:131–146

Furlan A (2012–2013) Il ruolo della struttura del paesaggio nella diversità floristica: il caso di studio della pianura veneta orientale. Tesi di Laurea

Gansi A (2007–2008) Analisi eco-idrologica della fascia di mobilita' laterale nel tratto potamale del fiume Oglio. Tesi di Laurea, Relatore: Prof. Pierluigi Viaroli

Gargano D, Passalacqua N, Bernardo L (2007) Bogs and Mires in Mediterranean Areas: The Vegetation of the Marshlands of Lacina Plain (Calabria, Italy). Phyton 47(1–2):161–189

Gasparri R, Casavecchia S, Galiè M, Biondi E (2013) The restoration of the wetlands with standing waters constituting the habitat of the Italian green toad (Bufo balearicus Boettger, 1880). Plant Sociol 50(1):109–119

Géhu JM, Biondi E (1988) Données sur la végétation des ceintures d'atterissement des lacs Alimini (Salento Italie). Doc Phytosoc 11:353–380

Géhu JM, Biondi E (1996) Synoptique des associations végétales du littoral adriatique italien. Giorn Bot Ital 130(1):257–270

Géhu JM, Costa M, Scoppola A, Biondi E, Marchiori S, Peris JB, Franck J, Caniglia G, Veri L (1984) Essai synsystématique et synchorologique sur les végétations littorales italiennes dans un but conservatoire. I - Dunes et vases salees. Doc Phytosoc 8:393–474

Gentile S, Guido MA, Montanari C, Paola G, Morucchio GB, Petrillo M (1988) Ricerche geobotaniche e saggi di cartografi a della vegetazione del piccolo bacino di Lago Riane (Liguria). Braun-Blanquetia 2:77–104

Gerdol R, Tomaselli M (1988) Phytosociology and ecology of stream vegetation in the summit region of the northern Apennines. Boll Museo Stor Nat Lunigiana 6:89–93

Gerdol R (1980) Flora e vegetazione della Palù Longa (Bolzano). Studi Trent Sci Nat Acta Biol 57:33–53

Gerdol R (1987) Geobotanical investigations in the small lakes of Lombardy. Atti Ist Bot Lab Critt Univ Pavia 7(6):5–49

Gerdol R, Bragazza L (2001) Syntaxonomy and community ecology of mires in the Rhaetian Alps (Italy). Phytocoenologia 31(2):271–299

Gerdol R, Piccoli F (1984) La vegetazione del Delta Padano: stato attuale delle conoscenze. Atti Soc Ital Sci Nat Milano 125:233–244

Gerdol R, Tomaselli M (1993) The vegetation of wetlands in the Northern Apennines (Italy). Phytocoenologia 21:421–469

Gerdol R, Tomaselli M (1997) Vegetation of wetlands in the Dolomites. Dissertationes botanicae, 281, Cramer, 197 pp

Giau M (1986) Indagine floristica e vegetazionale sullo stagno di Platamona (Sardegna Nord-occidentale). Boll Soc Sarda Sci Nat 25(1986):97–123. ISSN 0392-6710

Gigante D, Acosta ATR, Agrillo E, Attorre F, Cambria VM, Casavecchia S, Chiarucci A, Del Vico E, De Sanctis M, Facioni L, Geri F, Guarino R, Landi S, Landucci F, Lucarini D, Panfili E, Pesaresi S, Prisco I, Rosati L, Spada F, Venanzoni R (2012) VegItaly: Technical features, crucial issues and some solutions. Plant Sociol 49(2):71–79

Gigante D, Venanzoni R (2004) Flora e Vegetazione. In: AA.VV., Relazione sullo stato dell'Ambiente in Umbria, ARPA, Regione dell'Umbria, Perugia, pp 305–334

Giglio E, Tammaro F (1995) Compendio delle unità vegetazionali in Abruzzo. Atti dei Convegni Lincei 115:255–284

Granetti B (1965) La flora e la vegetazione del Lago trasimeno. Parte I: La vegetazione litoranea. Riv Idrobiol 4(3):115–153

Gubellini L, Hofmann H, Pinzi M (2014) Contributo alla conoscenza della flora vascolare delle Marche e di alcune regioni limitrofe. Inf Bot Ital 46(1):17–26

Guglielmetto Mugion L, Montacchini F (1993–94) La vegetazione del Lago di Viverone. Allionia 32:1–26

Guido M, Montanari C (1983) Studio e cartografia della vegetazione cacuminale del Monte Aiona (Appennino ligure). Arch Bot Biogeogr Ital 59(3–4):105–131

Iberite M, Paolozzi AM, Resini AM (1995) La vegetazione del lago di Bolsena (Viterbo, Italia centrale). Fitosociologia 29:151–164

ISPRA (2008) Repertorio nazionale degli interventi di ripristino. 6e Puglia Sistemazioni idraulico forestali a Marina di Lesina (documento on-line)

Landi M, Angiolini C (2006) the vegetation gradient alongthe longitudinal profile of a braided strem: a case study in central Italy. Belg J Bot 139(2):220–232

Landi M, Angiolini C, De Dominicis V (2002) Analisi fitosociologica dei fi umi della Toscana meridionale: il tratto medio-basso del Merse (Italia centrale). Stud Bot 21:37–88

Landucci F, Acosta ATR, Agrillo E, Attorre F, Biondi E, Cambria VM, Chiarucci A, Del Vico E, De Sanctis M, Facioni L, Geri F, Gigante D, Guarino R, Landi S, Lucarini D, Panfili E, Pesaresi S, Prisco I, Rosati L, Spada F, Venanzoni R (2012) VegItaly: the Italian collaborative project for a national vegetation database. Plant Biosyst 146(4):756–763

Landucci F, Gigante D, Venanzoni R, Chytrý M (2013) Wetland vegetation of the class Phragmito-Magno-Caricetea in central Italy. Phytocoenologia 43(1–2):67–100

Lasén C, Argenti C (1996) Due notevoli torbiere del Bellunese: Pra' Torond e Sochieva. Ann Mus Civ Rovereto 11:257–290

Lastrucci L, Becattini R (2008) La vegetazione delle aree umide presso Bosco ai Frati (Firenze, Toscana). Atti Soc Tosc Nat Mem Ser B 115:57–67

Lastrucci L, Foggi B, Gonnelli V, Gusmeroli E (2005) La vegetazione delle aree umide dei substrati ultramafici dell'Alta Valtiberina (Arezzo, Italia Centrale). Stud Bot 24:9–44

Lastrucci L, Foggi B, Selvi F, Becattini R (2007) Contributo alla conoscenza della vegetazione e della flora delle aree umide nel comprensorio di Capalbio (Provincia di Grosseto, Italia Centrale). Arch Geobot 10(1–2) (2004a): 1–30

Lastrucci L, Gonnelli V, Foggi B (2004) Flora e vegetazione di alcune aree umide dell'altopiano della "Pianca" nell'alta Val Marecchia (Provincia di Arezzo, Toscana). Inf Bot Ital 36:429–442

Lastrucci L, Landi M, Angiolini C (2010a) Vegetation analysis on wetlands in Tuscan agricultural landscape (Central Italy). Biologia 65:54–68

Lastrucci L, Landucci F, Gonnelli V, Barocco R, Foggi B, Venanzoni R (2012) The vegetation of the upper and middle River Tiber (Central Italy). Plant Sociol 49 (2):29–48

astrucci L, Paci F, Raffaelli M (2010b) The wetland vegetation of the Natural Reserves and neighbouring stretches of the Arno river in the Arezzo province (Tuscany, Central Italy). Fitosociologia 47:29–59

Lastrucci L, Viciani D, Nuccio C, Melillo C (2008) Indagine vegetazionale su alcuni laghi di origine artificiale limitrofi al Padule di Fucecchio (Toscana, Italia Centrale). Ann Mus Civ Rovereto Archeol Stor Sci Nat 23(2007):169–203

Lausi D, Gerdol R (1980) Mappe della vegetazione degli ambienti umidi subalpini delle Alpi Giulie occidentali. Friuli-Venezia Giulia (Provincia di Udine). Consiglio nazionale delle ricerche, Roma

Lonati M, Lonati S (2005) Le comunità a *Carex elata* All. della torbiera di Vanzone (Piemonte, Vercelli). Fitosociologia 42(2):15–21

Marchiori S, Sburlino G (1986) La vegetazione della palude Brusà (Cerea - Verona). Boll Mus Civ Sci Nat Verona 13:133–145

Marchiori S, Sburlino G (1997) Present vegetation of the Venetian Plain Allionia 34 (1996):165–180

Marchiori S, Sburlino G, Sillani L (1983a) Contributo alla conoscenza della flora e della vegetazione dei "Quadri" di Fagagna (UD). Atti Mus Civ Stor Na. Trieste Pag:65–79

Marchiori S, Sburlino G, Sillani L (1983b) Contributo alla conoscenza vegetazionale degli ambienti umidi della Pianura padana orientale: Valdorsa (VI). Giorn Bot Ital, 117(suppl. 1):127–128

Marchiori S, Sburlino G, Tisi F (1997) La vegetazione dei laghetti di Marco (Rovereto). Ann Mus Civ Rovereto 3:97–206

Marchiori S, Sburlino G, Tornadore N (1993) Check-list of the hydro-hygrophious vegetation of the Venetian plain. Giorn Bot Ital 127:720

Marchiori S, Tornadore Marchiori N, Reffo T (1980) La vegetazione della palude relitta di Onara (Tombolo, PD) – 1° contributo. Atti V Convegno Gr. "G. Gadio". Varese, maggio, pp 117–128

Mariotti MG, Braggio Morucchio G, Cornara L, Placereani S (1992) Etude phytosociologique et palynologique de la végétation actuelle et passée à Torre Guaceto (Pouilles, Italie méridionale). Candollea 47:31–60. En italien, résumés français et anglais

Mariotti MG (1995) Osservazioni sulla vegetazione della Liguria Atti Conv Lin 115:189–227

Martini F, Poldini L (1981) Il paesaggio vegetale del Fiume Noncello nell'area urbana di Pordenone. Gortania 2:123–156

Merloni N, Piccoli F (2001) La vegetazione del complesso Punte alberate e Valle Mandriole (Parco Regionale del Delta del Po - Italia). Braun-Blanquetia 29:1–17

Minghetti P, Pedrotti F (2000) La vegetazione del Laghetto delle Regole di Castelfondo (Trento). Studi Trent Sci Nat Acta Biol 74:175–189

Minissale P, Spampinato G (1990) Osservazioni fitosociologiche sul "Pantano Gurna" presso presso Mascali (Sicilia orientale). Boll Acc Gioenia Sci Nat 23 (336):317–336

Minissale P, Spampinato G (1995) L'impatto antropico sulla vegetazione della Riserva naturale orientata Fiume Fiumefreddo (Sicilia orientale). Coll Phytosoc 21:343–356

Minissale P, Spampinato G, (1986) Osservazioni fitosociologiche sul "Lago Gurrida" (Sicilia Nord-Orientale). Giorn Bot Ital 119(3–4)(1985):197–225

Montanari C, Guido MA (1980) La vegetazione idro-igrofila di alcune conche lacustri del versante nord di Monte Ragola (alta Val Nure - Appennino ligure-piacentino). Arch Bot Biogeogr Ital 56:13–42

Mossa L, Aru A, Fogu MC, Guarino R, Zavattero L (2008) Studio geobotanico del Parco Eolico di Ulassai

Orsomando E, Catorci A (1999) Carta della vegetazione naturale potenziale dell'Umbria (scala 1: 200.000). Note Esplicative. Reg. Umbria. Direzione Regionale Politiche Territoriali. Ambiente ed Infrastrutture. Dip. di Botanica ed Ecologia, Univ. Camerino S.EL.CA., Firenze

Orsomando E, Pedrotti F (1986) Le 'praterie galleggianti' a *Carex pseudocyperus* L. di alcuni laghi dell'Italia centrale. Riv Idrobiol 25(1–3):87–103

Panfili E, Pesciaioli P, Venanzoni R (2004) An Archive: un software per la gestione di dati d'erbario, floristici e vegetazionali da internet. Inf Bot Ital 36:110–112

Pedrotti F (1981–82) Il Caricetum lasiocarpae Koch 1926 del Lago Pudro in Trentino. Delpinoa 23–24:293–305

Pedrotti F (2015) A new plant association of Caricion gracilis alliance of the central Apennines (Italy). Contrib Bot 50:137–142

Pedrotti F (2016a) The caricetum distichae association in Italy. In: Box EO (ed) Vegetation structure and function at multiple spatial, temporal and conceptual scales, Geobotany Studies. Springer, Cham, pp 357–362

Pedrotti F (2016) ERRATUM. A new plant association of *Caricion gracilis* alliance of the central Apennines (Italy). Contrib Bot 51:155

Pedrotti (2014) In stampa

Pedrotti F, Cortini Pedrotti C, Dell'Uomo A (1969) Fitocenosi reofile del fiume Potenza. Giorn Bot Ital 103:626

Pedrotti F (1963) Nota sulla vegetazione steppica (Stipeto- Poion xerophilae e Diplachnion) nei dintorni di Trento. Studi Trent Sci Nat Acta Biol 40:288–301

Pedrotti F (1965) Censimento di ambienti umidi meritevoli di protezione: i piani carsici dell'Appennino umbro-marchigiano. Atti Ist Bot Lab Critt Univ Pavia 6:142–158

Pedrotti F (1967) Carta fitosociologica (1:3.000) della vegetazione dei Piani di Montelago (Camerino). Not Fitosoc 4:1–8

Pedrotti F (1976) Les prairies permanentes humides de l'Apennin Central: Phytosociologie et eartographie. Coll Phytosoc 5:181–187

Pedrotti F (1980) Guida all'escursione della Società Botanica Italiana in Val d'Adige e nel Parco Nazionale dello Stelvio. Univ. Camerino, Camerino, 27–30 Giugno 1980, pp 1–63

Pedrotti F (1981–82) Il Caricetum lasiocarpae Koch 1926 del Lago Pudro in Trentino. Delpinoa 23–24:293–305

Pedrotti F (1981c) Carte de la végétation de la région des Marches (Italie centrale) a l'échelle du 1/300.000. Doc Cart Ecol Grenoble 24:15–16

Pedrotti F (1982) La végétation du Pian Grande. Guide-Itinéraire, Excursion Internationale de Phytosociologie en Italie centrale, Univ. Camerino, Camerino, 2–11 juillet 1982, pp 347–360

Pedrotti F (1983) Alcuni ambienti umidi del Molise. Giorn Bot Ital 117(Suppl 1):131–132

Pedrotti F (1984) Piante rare e notevoli di alcuni ambienti umidi del Trentino. Atti Acc Roveretana Agiati VI(23):131–140

Pedrotti F (1985) Die Vegetation des Palù Redont (Trentino). Tuexenia 5:211–215

Pedrotti F (1986) La Palù di Borghetto in Valle Lagarina (Trentino). Inf Bot Ital 18(1–3):139–144

Pedrotti F (1988) La flora e la vegètazione del Lago di Loppio (Trentino). Giorn Bot Ital 122(3–4):105–147

Pedrotti F (1990) Stato dell'ambiente dei Laghi di Caldonazzo e Levico (Trentino) attraverso l'analisi della flora e vegetazione. Giorn Bot Ital 124:155

Pedrotti F (1991) Nota sulla flora e vegetazione del Lago di Madrano (Trentino). Inf Bot Ital 22:182–193

Pedrotti F (1995) Nota sulla.vegetazione degli ambienti umidi della bassa Valsugana (Trentino). Doc Phytosoc XV:417–449

Pedrotti F (2001) Environmental systems, vegetation belts and potential vegetation of the Monti Sibillini (Central Italy). Oecol Mont 10:13–18

Pedrotti F (2003) Il lago e il suo biotopo. In: Andreotti G, Cainelli G, Maschio R (eds) Atti Conv.: Il lago della Serraia: verso il suo recupero, Baselega di Piné 18/05/02, pp 41–44

Pedrotti F (2004) Ricerche geobotaniche al Laghestel di Piné (1967–2001). Braun-Blanquetia 35:1–54

Pedrotti F (2004) Vegetation mapping in wetlands. Ann Bot IV:29–36

Pedrotti F (2008) La vegetazione delle marcite di Norcia (Italia centrale). Braun-Blanquetia 44:1–31

Pedrotti F (1995) Flora e vegetazione del biotopo Lago di Toblino. Dattiloscritto, Trento

Pedrotti F, Chemini C (1981) La vegetazione del Laghestel di Piné (Trento). Studi Trent Sci Nat Acta Biol 58:425–462

Pedrotti F, Cortini Pedrotti C, Orsomando E (1979) The phytosociological map of Burano (Tuscany). Webbia 34(1):529–531

Pedrotti F, Cortini Pedrotti C (1982) Les "cariceta de Pian Perduto di Gualdo". Guide-Itinéraire, Excursion Internationale de Phytosociologie en Italie centrale, Univ. Camerino, Camerino, 2–11 juillet 1982, pp 337–340

Pedrotti F, Cortini Pedrotti C, Russo G (2003) The flora and vegetation of springs and waterways of the Gargano Promontory (Southern Italy). In: Abstracts 46th Symposium of I.A.V.S., Università degli Studi, Camerino, (8–14 June 2003 Napoli, Italia), p 60

Pedrotti F, Gafta D (1994) La palude di Roncegno. In: Pedrotti F (ed) Guida all'escursione della Società Italiana di Fitosociologia in Trentino (1–5 luglio 1994). Camerino, Dipartimento Botanica Ecologia, pp 123–129

Pedrotti F, Gafta D (1996) Ecologia delle foreste ripariali e paludose dell'Italia. L'Uomo e l'Ambiente 23:1–165

Pedrotti F, Gafta D (1992) Phytosociological and ecological research in a protected area as basis for its management. The example of Loppio Lake (North Italy). In: Applied vegetation ecology. I.A.V.S., Shangai, pp 148–150

Pedrotti F, Gafta D, Manzi A, Canullo R (1992) Le associazioni vegetali della piana di Pescasseroli (Parco Nazionale d'Abruzzo). Doc Phytos 14:123–147

Pedrotti F, Minghetti P (1994) Le Marocche di Dro. In: Pedrotti F (ed) Guida all'escursione della Società Italiana di Fitosociologia in Trentino (1–5 luglio 1994). Dipartimento di Botanica ed Ecologia, Camerino, pp 29–65

Pedrotti F, Minghetti P, Sartori G (1995) Evoluzione della vegetazione e del suolo delle marocche di Dro (Trento, Italia). Coll Phytosoc XXIV:203–222

Pedrotti F, Orsomando E (1979) Indagine geobotanica sul bacino del Trasimeno per la tutela e la valorizzazione del patrimonio naturalistico. Nuovo Giorn Bot Ital 113(5–6):458–459

Pedrotti F, Orsomando E (1982) Flore et vegetation du Lac Trasimene. In: Pedrotti F (ed) Guide-Itineraire de l'Excursion Internationale de Phytosociologie en Italie centrale. Università di Camerino, Camerino, pp 469–478

Pedrotti F, Orsomando E, Cortini Pedrotti C (1974) Carta della vegetazione del Parco nazionale dello Stelvio (Notizia esplicativa). Amm Parco Naz Stelvio:1–86

Pedrotti F, Pettorossi L (1969) Rilevamento cartografico della vegetazione della palude di Colfiorito. Mitt ostalp din pflanzensoz Arbeitsgem 9:153–159

Pedrotti F, Sanesi G (1969) Resoconto delle escursioni sull'Appennino Umbro-Marchigiano (25–27 giugno 1968). Mitt ostalp din pflanzensoz Arbeitsgem 9:365–388

Pedrotti F, Taffetani F (1982) La végétation des sources de Pontile et Fiuminata. Guide-Itineraire, Excursion Internationale de Phytosociologie en Italie centrale, Camerino, Univ. Camerino, 2–11 juillet 1982, pp 237–242

Pedrotti F, Venanzoni R (1987) La vegetazione di un bacino glacio-carsico sull'altipiano di Folgaria (Trentino). Webbia 41:289–304

Pellizzari M, Barbieri C, Caramori G, Pagnoni GA, Piccoli F (2007) La vegetazione della Salina di Comacchio (Ferrara, Parco del Delta del Po): ripristino ecologico

Pellizzari M, Merloni N, Piccoli F (1998) Vegetazione alonitrofila perenne nel Parco del Delta del Po (Ord. Juncetalia maritimi, All. *Elytrigio athericae – Artemision coerulescentis*). XXVIII Coll.e phytosoc. Camerino, 26–30 settembre: 138

Pellizzari M, Piccoli F (2001) La vegetazione dei corpi idrici del Bosco della Mesola (Delta del Po). Quad Staz Ecol Civ Mus St Nat Ferrara 13:7–24

Pellizzari M, Piubello F, Fogli S (2005) Aspetti vegetazionali del biotopo "Brusà-Vallette" (Cerea-Verona) e proposte per la conservazione degli habitat. Quaderni della stazione di ecologia del Civico Museo di Storia Naturale di Ferrara

Piccoli F, Merloni N (1989) Vegetation dynamics in coastal wetlands. An example in Northern Italy: the Bardello. Ecol Mediter XV, 112

Piccoli F (1976) Flora e vegetazione delle casse di espansione di Campotto e Valle Santa (Ferrara) (Flora and vegetation in the flood overflow areas of Campotto and Santa Valley), vol IV. Lavori della Società Italiana di Biogeografia, nuova serie

Piccoli F (1998) Flora e vegetazione. Analisi delle specie e delle comunità. In: Tinarelli R, Tosetti T (eds) Zone umide della pianura bolognese. I.B.C. Emilia-Romagna. Ed. Compositori, Bologna, pp 42–51

Piccoli F, Cortiselli S, Dell'Aquila L, Merloni N, Pellizzari M (1996) Vegetation map of the Regional Park of the Po Delta (Emilia - Romagna Region). Allionia 34:325–331

Piccoli F, Gerdol R (1979) La vegetazione di alcune zone umide del bolognese. Istituto ed orto Botanico Università di Ferrara, pp 83–97

Piccoli F, Gerdol R (1982) Rice field weed communities in Ferrara Province (Northern Italy). Aquat Bot 10:317–328

Piccoli F, Merloni N (1989) Vegetation dynamics in coastal wetlands, An example in Northern Italy: the Bardello. Ecol Mediterr 15:81–95

Piccoli F, Merloni N, Pellizzari M (1994) The vegetation of the Comacchio Saltern (Northern Italy). Ecol Mediter 20(3/4):85–94

Piccoli F, Pellizzari M, Dell'Aquila L (1999) Carta della vegetazione del Parco Regionale del Delta del Po. Stazioni Centro Storico e Valli di Comacchio. Scala 1:35.000. Regione Emilia Romagna, Servizio Cartografico e Geologico

Pignatti S (1953a) Introduzione allo studio fitosociologico della pianura veneta orientale. Atti Ist Bot Univ Lab Critt Pavia 11:92–258

Pignatti S (2005) Valori di bioindicazione delle piante vascolari della Flora Italiana. Braun-Blanquetia 39:1–97

Pignotti L (2003) *Scirpus* L. and related genera (*Cyperaceae*) in Italy. Webbia 58:281–400

Pirola A (1968) Appunti sulla vegetazione dei meandri del Ticino. Not Fitosoc 5:1–23

Pirone e Corbetta (1998)

Pirone G (1987) I magnocariceti degli Altipiani Maggiori d'Abruzzo. Inf Bot Ital 19:131–135

Pirone G (1991) Flora e vegetazione del Fiume Saline (Abruzzo). Micol Veg Mediterr 6(1):45–76

Pirone G (1997) Il paesaggio vegetale di Rivisondoli aspetti della flora e della vegetazione. Azienda Autonoma di Soggiorno e Turismo

Pirone G (2000) La vegetazione ripariale nei versanti nord-orientali del Gran Sasso d'Italiae dei Monti della Laga (Abruzzo, Italia). Fitosociologia 37(2):65–86

Pirone G, Ciaschetti G, Frattaroli AR, Corbetta F (2003) La vegetazione della Riserva Naturale Regionale Lago di Serranella (Abruzzo-Italia). Fitosociologia 40:55–71

Pirone G, Frattaroli AR, Corbetta F (1997) Vegetazione, cartografi a vegetazionale e lineamenti floristici della riserva naturale "Sorgenti della Pescara" (Abruzzo - Italia). Univ. degli St. dell'Aquila, Dip. di Scienze Ambientali, Popoli

Poldini L (1989a) La vegetazione acquatica, salmastra e palustre. In: Poldini L (ed) La vegetazione del Carso isontino e triestino. Lint, Trieste, pp 113–128

Poldini L (1989b) La vegetazione del Carso isontino e triestino. Lint, Trieste

Poli E, Maugeri G (1974) Studio della zonizzazione della vegetazione del Biviere. Arch Bot Biogeogr Ital 50(3–4):121–134

Presti G, Di Filippo C, Blasi C (2005) La vegetazione igrofila del Monumento Naturale Pantane e Lagusiello (Lazio centrale). Inf Bot Ital 36:401–408

Prosser F, Sarzo A (2003) Flora e vegetazione dei fossi nel settore trentino del fondovalle dell'Adige (Trentino-Italia settentrionale). Ann Mus Civ Rovereto Archeol Stor Sci Nat 18:89–144

Prosser F, Sarzo A (2004) L'area umida relitta di Roncafort: un biotopo di eccezionale interesse botanico (Trentino, Italia settentrionale). Ann Mus Civ Rovereto Archeol Stor Sci Nat 19:233–280

Provasi T (1926) Osservazioni e ricerche sulla vegetazione di alcuni laghetti dell'Appennino Tosco-Emiliano. Nuovo Giorn Bot Ital 33:681–725

Raffaelli M, Mori Secci M, Mariotti Lippi M, Fiorini G (1997) Indagini floristico-vegetazionali e actuopalinologiche sul lago Baccio e sul lago Greppo (Appennino Tosco-Emiliano). Webbia 51(2):201–235

Raimondo FM, Bazan G, Gianguzzi L, Ilardi V, Schicchi R, Surano N (2000) Carta del paesaggio e della biodiversità vegetale della Provincia di Palermo (Tav. 1: Balestrate-Partinico; Tav. 2: Palermo-Ustica; Tav. 3: Camporeale-Corleone; Tav. 4: Marineo-Caccamo; Tav. 5: Termini Imerese-Caltavuturo; Tav. 6: Cefalù-Petralia Sottana; Tav. 7: Contessa Entellina-Bisacquino; Tav. 8: Palazzo Adriano-Lercara Friddi; Tav. 9: Alia-Valledolmo; Tav. 10: Alimena-Gangi). Quad Bot Ambient Appl, 9 (1998). II: Allegati cartografici (Tav.1–10)

Reniero S, Cassol M, Ghirelli L, Scariot A, Vettorel S, Cailotto G, Crosara R, Dorigatti L (2013) WP4.2 – Pilot project PP7 "IT3250003 Biotopi litoranei della penisola del Cavallino", "IT3250013 Laguna del Mort e Pinete di Eraclea". Veneto Agricoltura

Rey C (1990) Flore et vegetation du Marais de Lozon (Verrayes-Vallée d'Aoste). Rev Valdôtaine d'hist Nat 44:5–21

Rey C, Desfayes M (1989) Un site marécageux de haut intérêt: le marais de Lozon. Rev Valdôtaine d'hist Nat 43:103–104

Rossi G, Alessandrini A (1998) Una banca dati sulla vegetazione delle aree protette in Emilia-Romagna. Arch Geobot 4:149–155

Sartori F (1980) Les forets alluviales de la basse vallee du tessin (Italie du nord). Colloques Phytosociologiques, IX, Les forêts alluviales, pp 201–216

Sartori F, Bracco F (1997) Present vegetation of the Po plain in Lombardy. Allionia 34:113–135

Sarzo A, Prosser F, Frisinghelli M (1997) Flora e vegetazione della zona umida di Bolzonella (Provincia di Padova – Italia settentrionale). Arch Geobot 3 (2):179–200

Sburlino G, Marchiori S (1987) La vegetazione idro-igrofila del medio corso del fiume Brenta (Veneto. Italia settentrionale). Univ de La Laguna Ser lnj 22:297–304

Sburlino G (1986) La vegetazione del Laghetto di Borgo Valsugana (TN). Studi Trent Sci Nat Acta Biol 62:113–123

Sburlino G, Bolzonella A, Marchiori S, Zanetti A (1989) Studi sulla vegetazione del Busatello (Veneto- Lombardia). 5. Contributo alla conoscenza della flora e della vegetazione. Boll Museo Civ Stor Nat Verona Bot Zool 7:33–46

Sburlino G, Buffa G, Filesi L, Gamper U, Ghirelli L (2013) Phytocoenotic diversity of the N-Adriatic coastal sand dunes - The herbaceous communities of the fixed dunes and the vegetation of the interdunal wetlands. Plant Sociol 50:57–77. (ISSN 2280-1855)

Sburlino G, Nicoletti F, Caniglia G (1995) La vegetazione acquatica e palustre delle cave di Gaggio nord (Marcon, Venezia). Soc Venez Sci Nat 20:133–145

Sburlino G, Tomaselli M, Oriolo G, Poldini L, Bracco F (2008) La vegetazione acquatica e palustre dell'Italia nord-orientale 2 - La classe Potametea Klika in Klika et V. Novák 1941. Fitosociologia 45(2):3–40

Scoppola A (1998) La vegetazione della Riserva Naturale Monte Rufeno (VT.). Reg. Lazio, Ass. U.T.V. delle risorse ambientali. Riserva Naturale Monte Rufeno, Comune di Acquapendente, Acquapendente, 69 pp

Scoppola A, Blasi C, Abbate G, Michetti L, Scagliusi E, Kuzminsky E, Antinori F (1990) La vegetazione della caldera del Lago di Vico. Regione Lazio, Univ. "La Sapienza", Roma

Scortegagna S, Tasinazzo S (2000) Flora dei Colli Berici (Vicenza - Italia settentrionale) I: Equisetaceae – Fabaceae. Nat Vicentina 4:69–119

Scortegagna S, Tasinazzo S (2001) Flora dei Colli Berici (Vicenza - Italia settentrionale) II: Oxalidaceae - Campanulaceae. Nat Vicentina 5:55–92

Scortegagna S, Tasinazzo S (2002) Flora dei Colli Berici (Vicenza - Italia settentrionale) III: Asteraceae - Orchidaceae. Nat Vicentina 6:87–121

Sortino M, Marceno C, Maggio F, Giannuzza A (1974) Tipologia e vegetazione riparia e lotica di due corsi d'acqua del versante nord del Fiume Platani (Sicilia centro-meridionale). Boll Studi Inf Giard Coloniale Palermo 24:72–102

Studio verde (2013) Piano di gestione del SIC IT6020012 "Piana di S. Vittorino – Sorgenti del Peschiera

Taffetani F, Rismondo M, Lancioni A (2011) Environmental evaluation and monitoring of agro-ecosystems biodiversity. In: Grillo O, Venora G (eds) Ecosystems biodiversity. InTech, Rijeka, pp 333–370

Tammaro F (1995) Lineamenti floristici e vegetazionali del Gran Sasso meridionale. Boll Museo Civ Stor Nat Verona 19:1–256

Tasinazzo S (2009) La vegetazione dei prati dei "Prà dei Gai" nella Pianura Veneta orientale (NE Italia). Fitosociologia 46(2):35–47

Tisi A, Minuzzo C, Siniscalco C, Caramiello C (2007) La vegetazione acquatica e palustre della zona dei "Cinque Laghi" di Ivrea. Riv Piem Stor Nat 28:87–126

Tomaselli M, Bernardo L (2006) La vegetazione degli ambienti umidi nei massicci del Sirino-Papa e del Pollino (Italia meridionale). Arch Geobot 9 (1–2) (2003):1–18

Tomaselli M, Bolpagni R, Gualmini M, Petraglia A, Longhi D (2006) Studio fitosociologico, cartografia della vegetazione ed analisi dello stato trofico delle acque delle "Paludi del Busatello" (Italia settentrionale). Boll Museo Civ Stor Nat Verona Bot Zool 30:3–37

Tomaselli M, Gualmini M, Petraglia A, De Santis S (2002) La Flora e la Vegetazione degli ambienti umidi dell'alta val parma e val cedra nel parco regionale dei cento Laghi (appennino parmense)

Tomaselli M, Manzini L, Del Prete C (1994) Carta della vegetazione con itinerari naturalistici 1:25.000 del Parco Regionale dell'Alto Appennino Modenese. Foglio Est. Servizio Cartografi co-Provincia di Modena, Settore Difesa del Suolo, Arti Grafi che Salomone, Roma

Tomaselli V (2001) Contributo alla conoscenza della vegetazione ripariale della Sicilia sud-orientale. Arch Geobot 7(2):11–24

Tomaselli V, Di Pietro R, Sciandrello S (2011) Plant communities structure and composition in three coastal wetlands in southern Apulia (Italy). Biologia 66 (6):1027–1043

Tomaselli V, Sciandrello S, Dibitonto P, Wagensommer RP, Urbano M, Calabrese IT, Garziano G, Cimmarusti G, Di Pietro R (2010) Analisi del paesaggio vegetale ed agricolo della Riserva Naturale Statale di "Torre Guaceto" (Brindisi - Puglia). Cartografia della vegetazione, degli habitat e dell'uso del suolo. Quad Bot Amb Appl 21(2010):33–49

Tomasi D, Caniglia G (2004) La vegetazione de le Poscole: importante sito di biodiversit^ in Provincia di Vicenza (N-E Italia). Lav Soc Ven Sci Nat 29:71–78

Tomasi D, Dal Lago A, Caniglia G (2004) "Le Poscole": importante fonte di biodiversità in provincia di Vicenza (Italia nordorientale). Studi Trent Sci Nat Acta Biol 80:245–246

Tomei PE, Guazzi E, Kugler PC (2001) Le zone umide della Toscana: indagine sulle componenti floristiche e vegetazionali. Edizioni Regione Toscana, Toscana

Tomei PE, Guazzi E, Spinelli P (1998) La vegetazione delle zone umide della Toscana: prime considerazioni. Atti Mus Stor Nat Maremma 17:195–206

Vagge I, Biondi E (1999) La vegetazione delle coste sabbiose del Tirreno settentrionale italiano. Fitosociologia 36(2):61–95

Vanacore Falco I, Venanzoni R (2007–2008) Indagine floristico-vegetazionale nel Sito d'Importanza Comunitaria "Talweg della Val Ferret" (IT 1204032) (Courmayeur, Aosta). Rev Valdôtaine Hist Nat 63:23–62

Venanzoni e Gigante (2004)

Venanzoni (1984) Ricerche sull'ecologia di Carex gracilis Curt. nei Piani di Montelago. Giorn Bot Ital 118(suppl. 2):121

Venanzoni R (1984) Alcuni resti di vegetazione palustre e torbosa in Val di Vizze (Alto Adige). Studi Trentini Sci Nat Acta Biol 61:169–180

Venanzoni R (1988a) Contributo alla conoscenza dei prati umidi della Sila (Calabria - Italia). Doc Phytosoc 11:613–633

Venanzoni R (1988b) La vegetazione della torbiera «Pezzabosco» (Trentino orientale). Studi Trent Sci Nat Acta Biol 64(1987):95–113

Venanzoni R (1988c) Segnalazione dell'associazione *Valeriano-Filipenduletum* Siss. 1946 nell'alta Valle del Velino. Giorn Bot Ital 117(suppl. 1):133

Venanzoni R (1991) La vegetazione di alcune radure formatesi in seguito a schianti provocati dal vento in Valsugana (Trentino). Inf Bot Ital 22:194–196

Venanzoni R (1992) I prati umidi e inondati dell'Alta Valle del Velino (Rieti-Italia Centrale). Doc Phytosoc 14:149–164

Venanzoni R (1995) Flora e vegetazione del Biotopo "Lago di Cei": gli ambienti umidi. Studi Trent Sci Nat Acta Biol 70:77–98

Venanzoni R (2014) Flora e la vegetazione del Lago «Pudro» (Trentino). Studi Trent Sci Nat Acta Biol Trento. [manuscript]

Venanzoni R, Apruzzese A, Gigante D, Suanno G, Vale F (2003) Contributo alla conoscenza della vegetazione acquatica e igrofitica dei Laghi di Monticchio. Inf Bot Ital 35:69–80

Venanzoni R, Canullo C (1984) Accrescimento densità e biomassa di Carex gracilis Curt. nei Piani di Montelago (Appennino centrale). Inf Bot Ital 16(2–3):155–164

Venanzoni R, Gigante D (2000) Contributo alla conoscenza della vegetazione degli ambienti umidi dell'Umbria (Italia). Fitosociologia 37(2):13–63

Venanzoni R, Gigante D (2005) Lo studio della vegetazione in Umbria: fondamenti metodologici e settori di indagine. Inf Bot Ital 37: ISSN: 0020-0697

Venanzoni R, Gigante D, Montagnoli L, Frattegiani M (2006) Habitat e Specie della Direttiva 92/43/CEE ed altri aspetti di rilevanza naturalistica al Lago Trasimeno. Di AA. VV, AP&elle s.r.l., Perugia, p 96

Zanotti Censoni AL, Corbetta F (1981) Boschi igrofili ad *Alnus glutinosa* in Lomellina. Not Fitosoc 17:33–44

References

Angius R, Bacchetta G (2009) Boschi e Boscaglie ripariali del Sulcis-Iglesiente (Sardegna sud-occidentale - Italia). Braun Blanquetia 45:3–64

Aquaro G, Carapelli KF, Peruzzi L (2008) The genus *Taraxacum* (*Asteraceae*) in Italy. I. A sistematic study of *Taraxacum* sect. *Palustria*. Phytol Balc 14(1):61–67

Bacchetta G, Mossa L (2004) Studio fitosociologico delle cenosi a Carex microcarpa Bertol. ex Moris della Sardegna meridionale. Fitosociologia 41(1, suppl 1):171–178

Balátová-Tuláčková E (1963) Zur Systematik der europäischen Phragmitetea. Preslia 35:118–122

Balátová-Tuláčková E (1994) Magnocaricion elatae-Gesellschaften - Eine Ergänzung zum Werk "Die Pfl anzengesellschaften Österreichs". Verh Zool Bot Ges Österreich 131:27–36

Balátová-Tuláčková E, Pavlíček V (1980) Beitrag zur Kenntnis der chemischen Zusammensetzung der Phragmiti-Magnocaricetea und der Agropyro-Rumicion Arten. Folia Geobot Phytotax 15:225–244

Balátová-Tuláčková E, Mucina L, Ellmauer T, Wallnöfer S (1993) Phragmiti-Magnocaricetea. In: Grabherr G, Mucina L (eds) Die Pfl anzengesellschaften Österreichs, Teil 2, Natürliche waldfreie Vegetation. Gustav Fischer Verlag, Jena, pp 79–130

Bardat J, Bioret F, Botineau M, Boullet V, Delpech R, Géhu J-M, Haury J, Lacoste A, Rameau J-C, Royer J-M, Roux G,Touffet J (2001) Prodrome des Végétations de France // Version 01-2 [14 décembre 2001] // 1

Bardat J, Bioret F, Botineau M, Boullet V, Delpech R, Géhu J-M, Haury J, Lacoste A, Rameau J-C, Royer J-M, Roux G, Touffet J (2004) Prodrome des végétations de France. Coll Patrimoines naturels, 61. Muséum National d'Histoire Naturelle, Paris, 171 pp

Beaufoy G, Baldock D, Clark J (1994) The nature of farming: low intensity farming systems in nine European countries. IEEP, London

Bignal EM, McCraken DI (2000) The nature conservation value of European traditional farming systems. Environ Rev 8:149–171

Biondi E, Bracco F, Nola P (eds) (1996) Bibliografia fitosociologica italiana (1909–1993). Fitosociologia 31:5–10

Biondi E, Bracco F, Nola P (eds) (1997) Lista delle unità sintassonomiche della vegetazione italiana. Fitosociologia 33:5–11

Biondi E, Burrascano S, Casavecchia S, Copiz R, Del Vico E, Galdenzi D, Gigante D, Lasen C, Spampinato G, Venanzoni R, Zivkovic L, Blasi C (2012) Diagnosis and syntaxonomic interpretation of Annex I Habitats (Dir. 92/43/EEC) in Italy at the alliance level. Plant Sociol 49(1):5–37

Biondi E, Blasi C, Allegrezza M, Anzellotti I, Azzella MM, Carli E, Casavecchia S, Copiz R, Del Vico E, Facioni L, Galdenzi D, Gasparri R, Lasen C, Pesaresi S, Poldini L, Sburlino G, Taffetani F, Vagge I, Zitti S, Zivkovic L (2014) Plant communities of Italy: the vegetation prodrome. Plant Biosyst 148:728–814

Bracco F (2001) Il primo aggiornamento della Lista sintassonomica italiana. Fitosociologia 38 (2, Suppl. 1):5–11

Bracco F, Nola P (1995) Project LISY: data banks for the compilation of a national list of syntaxonomic units of Italy. Ann Bot (Roma) 53:163–169

Bracco F, Panfili E, Sburlino G, Venanzoni R (2007) Il secondo aggiornamento della Lista sintassonomica italiana (2000–2004). Fitosociologia 44(1, Suppl 1):5–12

Brullo S, Ronsisvalle GA (1975) La vegetazione dei Gorghi Tondi e del lago Preola presso Mazara del Vallo (Sicilia occidentale). Not Fitosoc 10:45–67

Chytrý M (ed) (2011) Vegetace České Republiky 3. Vodní a mokř adní vegetace. Vegetation of the Czech Republic 3. Aquatic and wetland vegetation. Academia, Praha

Conti F, Abbate G, Alessandrini A, Blasi C (eds) (2005) An annotated check-list of the Italian vascular flora. Palombi, Roma

Cortini PC, Orsomando E, Pedrotti F, Sanesi G (1973) La vegetazione e i suoli del Pian Grande di Castelluccio di Norcia (Appennino centrale). Atti Ist Bot Lab Critt Univ Pavia, 1973, IX(serie 6):155–249

Costalonga S, Venanzoni R (2009) Notula: 1600. Inf Bot Ital 41(2):343–361

Dunkel FG (2011) The Ranunculus auricomus L. complex (Ranunculaceae) in Central and Southern Italy, with additions to North Italian taxa. Webbia 66(2):165–193

Foggi B, Cartei L, Pignotti L, Signorini MA, Viciani D, Dell'Olmo L, Menicagli E (2006) Il paesaggio vegetale dell'Isola d'Elba (Arcipelago Toscano). Studio fitosociologico e cartografico. Fitosociologia 43(suppl 1):3–95

Gigante D, Acosta ATR, Agrillo E, Attorre F, Cambria VM, Casavecchia S, Chiarucci A, Del Vico E, De Sanctis M, Facioni L, Geri F, Guarino R, Landi S, Landucci F, Lucarini D, Panfili E, Pesaresi S, Prisco I, Rosati L, Spada F, Venanzoni R (2012) VegItaly: technical features, crucial issues and some solutions. Plant Sociol 49(2):71–79

Grabherr G, Mucina L (1993) Die Pflanzengesellschaften Österreichs, Teil II, Natürliche waldfreie Vegetation. Gustav Fisher Verlag, Jena

Hooper DU, Chapin FS, Ewel JJ et al (2005) Effects of biodiversity on ecosystem functioning: a consensus of current knowledge. Ecol Monogr 75:3–5

ISPRA (2011) Contributi per la tutela della biodiversità delle zone umide. Rapporti ISPRA n. 153/2011

IUCN (2008) http://www.iucn.org/news_homepage/news_by_date/2008_news/

Jiménez-Mejías P, Luceño M, Martín-Bravo S et al (2011) The importance of the Italian peninsula in the phylogeography of Carex (Cyperaceae). Inf Bot Ital 44(Suppl. 2):53

Jiménez-Mejías P, Martín-Bravo S, Amini-Rad M, Luceño M (2014) Disentangling the taxonomy of Carex acuta s.l. in the Mediterranean basin and Middle East: Re-evaluation of C. panormitana Guss. and C. kurdica Kük ex Hend.-Mazz. Plant Biosyst 148(1–2):64–73

Landucci F, Acosta ATR, Agrillo E, Attorre F, Biondi E, Cambria VM, Chiarucci A, Del Vico E, De Sanctis M, Facioni L, Geri F, Gigante D, Guarino R, Landi S, Lucarini D, Panfili E, Pesaresi S, Prisco I, Rosati L, Spada F, Venanzoni R (2012) VegItaly: the Italian collaborative project for a national vegetation database. Plant Biosyst 146(4):756–763

Landucci F, Gigante D, Venanzoni R, Chytrý M (2013) Wetland vegetation of the class Phragmito-Magno-Caricetea in central Italy. Phytocoenologia 43(1–2):67–100

Lucarini D, Gigante D, Landucci F, Panfili E, Venanzoni R (2015) The anArchive taxonomic checklist for Italian botanical data banking and vegetation analysis: theoretical basis and advantages. Plant Biosyst 149(6):958–965. doi:10.1080/11263504.2014.98401

Masci S, Miho A, Marchi P (1994) *Ranunculus auricomus* L. aggr. (Ranunculaceae) in Italy. I. Sexual tetraploids on the Apennines. Caryologia 47:97–108

Matulevičiūtė (2002) Diversity and distribution of communities of the Magnocaricetalia elatae Pignatti (1953) 1954 orderin Lithuania [Magnocaricetalia elatae Pignatti (1953) 1954 eilės bendrijų įvairovė ir paplitimas Lietuvoje]. Bot Lithuanica 8(1):3–31

Molina JA, Moreno PS (2003) Diversity of the helophytic vegetation in Spain. Acta Bot Gall 150:167–193

Mossa L, Aru A, Fogu MC, Guarino R, Zavattero L (2008) Studio geobotanico del Parco Eolico di Ulassai. Edibo, Catania, p 193

Ninot JM, Carreras J, Carrillo E, Vigo J (2000) Syntaxonomic conspectus of the vegetation of Catalonia and Andorra. I Hygrophylous Herbaceus communities. Acta Bot Barcinon 46:191–237

Oberdorfer E (1977) Suddeutsche PFL plfanzengesellschaften, vol 1. Fischer, Verlag

Oťahelová H, Hrivnák R, Valachovič M (2001) Phragmito-Magnocaricetea Klika in Klika et Novák 1941. In: Valachovič M (ed) Rastlinné spoločenstvá Slovenska. 3. Vegetácia mokradí. Veda, Bratislava, pp 53–183

Pedrotti F (1971) Carex buxbaumii, relitto glaciale: prima segnalazione per l'Italia peninsulare. Arch Bot Biogeogr Ital, XLVII, pp 21–28

Pedrotti F (1977) Sulla presenza di Carex disticha Huds. e di Carex repens Bell. in Italia Atti 1st Bot Lab Critt Univ Pavia, s. 6, XII, 147–161

Pedrotti F (1983) La conservazione degli ambienti umidi in Italia. Mem Soc Geogr Ital, XXXIII, pp 49–68

Pedrotti F (1985) Géomorphologie et répartition de la végétation dans les bassins karstiques des Apennins. Colloq Phytosoc XIII:507–539

Pedrotti F (2013) Ambienti residui e minacciati. In Bologna G (ed) Rapporto WWF, WWF report 2013, pp 118–122

Pedrotti F, Sanesi G (1969) Resoconto delle escursioni sull'Appennino Umbro-Marchigiano (25-27 giugno 1968). Mitt ostalp din pflanzensoz Arbeitsgem 1969(9):365–388

Peruzzi L, Carlesi V (2012) The genus Taraxacum (Asteraceae) in Italy. IV. Two new species of T. sect. Palustria. Willdenowia 42(2):191–197

Pignatti S (1953) Introduzione allo studio fitosociologico della pianura veneta orientale con particolare riguardo alla vegetazione litoranea. Arch Bot 29:65–98. 3 S, XII(2)

Postel S, Carpenter S (1997) Freshwater ecosystem services. In: Daily GC (ed) Nature's services: societal dependence on natural ecosystems. Island Press, Washington, pp 195–214

Pott R, Remy D (2000) Gewässer des Binnenlandes. Ulmer, Stuttgart

Rivas-Martínez S, Fernández-González F, Loidi J, Lousã M, Penas A (2001) Syntaxonomical checklist of vascular plant communities of Spain and Portugal to association level. Itinera Geobot 14:5–341

Rivas-Martínez S, Díaz TE, Fernández-González F, Izco J, Loidi J, Lousã M, Penas A (2002) Vascular plant communities of Spain and Portugal. Addenda to the Syntaxonomical checklist of 2001. Itinera Geobot 15(1–2):5–922

Schaminée JHJ, Stortelder AHF, Westhoff V (1995) De Vegetatie van Nederland. Opulus Press, Uppsala, SE

Taffetani F, Rismondo M, Lancioni A (2011) Environmental evaluation and monitoring of agro-ecosystems biodiversity. In: Grillo O, Venora G (eds) Ecosystems biodiversity. InTech, Rijeka, pp 333–370

Urbani M, Calvia G, Pisanu S (2013) *Carex panormitana* Guss. Inf Bot Ital 45(1):115–193

Valachovič M (ed) (2001) Rastlinné spoločenstvá Slovenska 3. Vegetácia mokradí. Veda, Bratislava

Venanzoni R, Gigante D (2000) Contributo alla conoscenza della vegetazione degli ambienti umidi dell'Umbria. Fitosociologia 37(2):13–63

Venanzoni R, Landucci F, Panfili E, Gigante D (2012) Toward the Italian national vegetation database: VegItaly. Biodivers Ecol 4:185–190

Weber HE, Moravec J, Theurillat JP (2000) International code of phytosociological nomenclature, 3rd edition. J Veg Sci 11:739–768

European Grasslands Gradient and the Resilience to Extreme Climate Events: The SIGNAL Project in Italy

9

Kevin Cianfaglione, Stefano Chelli, Giandiego Campetella, Camilla Wellstein, Marco Cervellini, Sandro Ballelli, Domenico Lucarini, Roberto Canullo, and Anke Jentsch

Abstract

Extreme weather events, land use, and the presence of invasive species can act as pressures threatening biodiversity, resilience and ecosystem services. Particularly in the open cultural landscape, these pressures can suddenly drive ecosystems across tipping points and beyond thresholds of system integrity. Yet, biodiversity holds features for buffering against change. Potential stabilizing mechanisms include species richness, presence of key species such as legumes, and intraspecific diversity. These potential buffers can be promoted by conservation management and political decisions. On this basis, the results of the SIGNAL project and the related scientific based knowledge, will help to develop recommendations that could be used by International, national and local authorities, EU institutions, NGOs, for agriculture, landscape management and biodiversity conservation. Here we briefly describe the project and the Italian field experimental site.

K. Cianfaglione (✉)
School of Biosciences and Veterinary Medicine - Plant Diversity and Ecosystems Management Unit, University of Camerino, Via Pontoni 5, 62032 Camerino, Italy

EA 2219 Géoarchitecture, UFR Sciences & Techniques, Université de Bretagne Occidentale, 6 Avenue Victor Le Gorgeu, 29200 Brest, France
e-mail: kevin.cianfaglione@unicam.it

S. Chelli • G. Campetella • M. Cervellini • S. Ballelli • D. Lucarini • R. Canullo
School of Biosciences and Veterinary Medicine - Plant Diversity and Ecosystems Management Unit, University of Camerino, Via Pontoni 5, 62032 Camerino, Italy

C. Wellstein
Faculty of Sciences and Technology, Free University of Bolzano/Bozen, Piazza Università 1, 39100 Bolzano, Italy

A. Jentsch
Disturbance Ecology, University of Bayreuth, Universitätsstr. 30, 95447 Bayreuth, Germany

F. Pedrotti (ed.), *Climate Gradients and Biodiversity in Mountains of Italy*, Geobotany Studies, https://doi.org/10.1007/978-3-319-67967-9_9

Keywords
Biomass • Climate change • Disturbance • Diversity • Experiment • Extreme event • Gradient • Grasslands • Invaders • Management • Manipulation • Meadows • Mowing • Multifunctional • Value

9.1 Introduction

European grasslands provide many ecosystem services and have a crucial influence on the landscape and the quality of life. Extreme weather events (e.g.: extreme drought), changes in the land use, and the presence of invasive alien species, can threaten biodiversity, resilience and the services provided by these systems (see: Leadley et al. 2010). On the other hand, the complexity of the landscape mosaic and the biodiversity of these phytocoenoses can act as a buffer to such pressures, like a "Shock absorber" (*sensu* Maynard-Smith 1974; de Santo 1978). The resilience and potential mechanisms of stabilization are represented by some factors, like species richness (insurance hypothesis, Yachi and Loreau 1999), presence of key species (Spehn et al. 2002) and intraspecific diversity. Studies including both species richness and traits have been claimed as a promising way of addressing ecosystem processes (Díaz and Cabido 2001).

The SIGNAL project will investigate the major mechanisms of resilience in European grasslands under simulated extreme drought events in order to identify early warning signals of thresholds. The results will provide experimental evidence, which may serve to regulate ecosystem services and foster the identification of early warning signals, including intraspecific and functional groups diversity. Moreover, the influence of different mowing regimes and of potentially invasive plant on grassland diversity and resilience are tested. SIGNAL will intensely serve the science/policy interface by delivering highly relevant and understandable information to policy makers and stakeholders about European gradients of resilience and indicators of tipping points in the face of climate extremes, supporting EU policies and national legislative frameworks (Campetella et al. 2013; Cianfaglione et al. 2013; Wellstein et al. 2014a; Jentsch et al. 2014). In this paper, we briefly describe the Signal project, with particular regard to the Italian experimental activities.

9.2 The Signal Project

The SIGNAL research project was funded within "BiodivERsA" call, under the FP7 scheme "ERA-Net". The research project started in March 2013; with the duration of 3 years.

SIGNAL questions whether elements of biodiversity stabilizing mechanisms in the face of extreme weather events such as drought can be promoted by changes in land use techniques. It is aimed at identifying early warning signals of change and providing support in resilience management through conservation management and land use—including altered land management practices. By newly implementing a coordinated distributed field experiment, three hypotheses have been set as follows: extreme weather events (H1), land-use changes (H2) and the presence of invasive species (H3) can act as pressures threatening biodiversity, resilience and ecosystem services of semi-natural grasslands and can suddenly drive them beyond thresholds of system integrity (tipping points and regime shift). On the other hand, biodiversity itself may buffer against change. Potential stabilizing mechanisms include species richness, presence of key species such as legumes and within species diversity (Jentsch et al. 2014). On this basis, scientific knowledge based recommendations will be developed, that could be used by a range of stakeholders (International, national and local authorities, non-governmental organizations) for agriculture, landscape, biodiversity management and conservation, as well as for livestock and forestry purposes.

9.3 The Study Area

Ten countries join the project with respective experimental fields (Belgium, Bulgaria, France, Germany, Hungary, Italy, Switzerland, Turkey, Israel, Austria). The sites differ in climate conditions, management, disturbance history, geology and edaphic factors, they are organized in a latitudinal-climatic gradient and represent common types of grasslands of the respective study country (Wellstein et al. 2014b). Our lawn, is a 700 m^2 (35×20 m) stable meadow in a flat area (Fig. 9.1), located in the "*Arboretum apenninicum*", established in 1990 by the Camerino University, in an area of 10 hectares located in Tuseggia di Camerino, Marche region (Central Italy), between 495 and 582 m a.s.l. (Pedrotti 1992). Referring to Camerino, the average yearly temperature is 12 °C and the annual average rainfall is 1100 mm. Our field instruments revealed a mean annual precipitation average 880 mm and a mean annual temperature of 12.1 °C. In 1990 the study area was already not cultivated for several years. It is a typical formation of secondary meadows, from low-altitude, in the hilly belt (*Quercus pubescens* zone); where the potential vegetation is mostly attributable to the *Peucedano cervariae-Quercetum pubescentis*, often interspersed by other residual fragments and by several associations in substitution (Cianfaglione 2014). The geologic substrate is the classic Flysh from the Marche hills, with alternating layers of sandstone, clay or marl; the soil is deep, with a silty-clay or clay texture. These types of meadow are often characterized by different plant communities and plant groupings; often less studied. In particular, our meadow consists of some small patches characterized by more or less nitrophilous, humid, ruderal, disturbed, or trampled conditions. Dry local conditions are related to disturbance and ruderality.

During centuries, these fields often represent areas alternately undergone various land uses depending on the occurrences and on the crop rotation management. This stratification of uses over the time, make a layer combination, of "Fossil Landscape" traces (*sensu* Gorfer 1982). The main landscape elements of the area are: arable fields, meadows, small villages, vineyard, farmhouses, woodlands, coniferous

Fig. 9.1 Meadow from the Italian site of the SIGNAL experiments, and its surroundings; in spring habitus. In the background, some little Pedunculate Oaks in the *Quercus pubescens* belt

reforestation for slopes hydrogeological protection, shrub land, bocages and orchards (including olive and chestnut plantations). This context represents the typical hilly mosaic vegetation, throughout Central Italy, characterizing the landscape in relatively small parcels used in a multifunctional way. This type of meadow is very important from the geography, ecology, agronomy, and landscape point of views. They are rich in plant species (and communities), and the composition is characterized by remarkable presence of officinal, foraging, endemic and segetal species. This reflects their polyvalent importance in botany, apiculture, zoology, esthetics, forestry and agronomy. They are also valuable for livestock farming, producing a lot of good quality biomass. Before the experiment, our studied meadow was normally mowed about 2–3 (exceptionally even 4) times per year. The presence of abandoned surfaces is important, for the Landscape mosaic heterogeneity, natural dynamics, biodiversity, and secondary succession. Global changes (i.e.: climate, land use, alien/native species ratio) can threaten the productivity, functionality and diversity of these formations. These meadow systems are characterized by a certain presence of exotic species due to their ruderal, anthropic and agricultural influences. Threats are primarily the intensification and the extension of arable lands, the landscape homogenization caused by the modern agricultural practices; the spread of crops that do not rotate sufficiently over the time; urbanization; forest logging intensification; climate changes; and possible new alien species problems.

Our study field contained 162 species (Table 9.1) with a record of 14 species (overlapping + rooting) casually found during a 5 × 5 cm quadrat survey. The maximum number of rooting species was 12, found casually in another similar quadrat.

Critical specimen have been collected, classified and stored at the "*Herbarium Camerinenis*" (CAME), University of Camerino.

Table 9.1 Species list and general cover of the Italian site. Ligneous species are here only revealed as saplings or seedlings

n.	Species list	Cover x > 10%; r = rare
1	*Acer campestre* L.	
2	*Acer pseudoplatanus* L.	
3	*Achillea millefolium* L.	x
4	*Acinos alpinus* (L.) Moench subsp. *alpinus*	
5	*Agrimonia eupatoria* L. subsp. *eupatoria*	
6	*Ajuga chamaepytis* (L.) Schreb. subsp. *chamaepytis*	r
7	*Ajuga reptans* L.	
8	*Allium vineale* L.	r
9	*Alopecurus myosuroides* Huds.	
10	*Althaea hirsuta* L.	r
11	*Anagallis arvensis* L. subsp. *arvensis*	
12	*Anisantha sterilis* (L.) Nevski	
13	*Anthemis cotula* L.	
14	*Arctium minus* (Hill) Bernh.	
15	*Arenaria leptoclados* (Rchb.) Guss.	
16	*Arrhenatherum elatius* (L.) P. Beauv. ex J. Presl et C. Presl subsp. *elatius*	x
17	*Arthemisia verlotiorum* Lamotte	
18	*Arum italicum* Mill. subsp. *italicum*	r
19	*Avena fatua* L. (group)	
20	*Bellevalia romana* (L.) Sweet	r
21	*Bellis perennis* L.	x
22	*Blackstonia perfoliata* (L.) Huds. subsp. *perfoliata*	r
23	*Brachypodium rupestre* (Host) Roem. et Schult. subsp. *rupestre*	
24	*Bromopsis erecta* (Huds.) Fourr. subsp. *erecta*	
25	*Bromus gussonei* Parl.	
26	*Bromus hordeaceus* L. subsp. *hordeaceus*	
27	*Bromus racemosus* L.	
28	*Calamintha nepeta* (L.) Savi subsp. *nepeta*	r
29	*Campanula rapunculus* L.	
30	*Capsella rubella* Reut.	
31	*Carduus pycnocephalus* L. subsp. *pycnocephalus*	
32	*Carex divulsa* Stokes	
33	*Carex flacca* Schreb. subsp. *erythrostachys* (Hoppe) Holub.	r
34	*Catapodium rigidum* (L.) C.E. Hubb. ex Dony subsp. *rigidum*	r
35	*Centaurea nigrescens* Willd. subsp. *neapolitana* (Boiss.) Dostál	
36	*Centaurium erythraea* Rafn subsp. *erythraea*	r
37	*Cephalaria transsylvanica* (L.) Roem. et Schult.	x
38	*Cerastium glomeratum* Thuill.	x
39	*Cerastium brachypetalum* subsp. *tenoreanum* (Ser.) Soó & Jáv., gen.	r

(continued)

Table 9.1 (continued)

n.	Species list	Cover x > 10%; r = rare
40	*Chenopodium album* L. subsp. *album*	r
41	*Cichorium intybus* L. subsp. *intybus*	x
42	*Cyclamen hederifolium* Aiton	r
43	*Cirsium vulgare* (Savi) Ten.	
44	*Clematis vitalba* L.	
45	*Colutea arborescens* L.	
46	*Conium maculatum* L.	
47	*Convolvulus arvensis* L.	
48	*Cota tinctoria* (L.) J. Gay subsp. *australis* (R. Fern.) Oberpr. et Greuter	
49	*Crataegus monogyna* Jacq.	
50	*Crepis pulchra* L.	
51	*Crepis sancta* (L.) Babc. subsp. *sancta*	
52	*Crepis vesicaria* L.	x
53	*Cruciata laevipes* Opiz	
54	*Dactylis glomerata* L. subsp. *glomerata*	x
55	*Daucus carota* L. subsp. *carota*	x
56	*Echium vulgare* L. subsp. *vulgare*	
57	*Elytrigia repens* (L.) Nevski	
58	*Epilobium tetragonum* L. (group)	
59	*Erigeron canadensis* L.	
60	*Erigeron sumatrensis* Retz.	
61	*Erodium malacoides* (L.) L'Hér. subsp. *malacoides*	
62	*Euphorbia helioscopia* L. subsp. *helioscopia*	
63	*Euphorbia peplus* L.	r
64	*Ficus carica* L.	r
65	*Galega officinalis* L.	
66	*Galium aparine* L.	
67	*Galium mollugo* L. subsp. *erectum* Syme	
68	*Geranium dissectum* L.	x
69	*Geranium pusillum* L.	r
70	*Himantoglossum adriaticum* H. Baumann	r
71	*Holcus lanatus* L.	
72	*Hypericum perforatum* L. subsp. *veronense* (Schrank) Ces.	
73	*Hypochaeris achyrophorus* L.	
74	*Inula conyzae* (Griess.) Meikle	
75	*Kickxia spuria* (L.) Dumort.	r
76	*Knautia integrifolia* (L.) Bertol. subsp. *integrifolia*	
77	*Lactuca saligna* L.	
78	*Lactuca serriola* L.	x
79	*Lamium maculatum* L.	
80	*Lamium purpureum* L.	

(continued)

Table 9.1 (continued)

n.	Species list	Cover x > 10%; r = rare
81	*Lapsana communis* L. subsp. *communis*	
82	*Lathyrus aphaca* L. subsp. *aphaca*	
83	*Lathyrus sphaericus* Retz.	
84	*Lathyrus sylvestris* L. subsp. *sylvestris*	
85	*Legousia speculus-veneris* (L.) Chaix	r
86	*Lepidium campestre* (L.) R. Br.	r
87	*Leucanthemum vulgare* (Vaill.) Lam. subsp. *vulgare*	
88	*Linaria vulgaris* Mill. subsp. *vulgaris*	
89	*Linum bienne* Mill.	
90	*Linum corymbulosum* Rchb.	
91	*Lolium perenne* L.	
92	*Malva sylvestris* L. subsp. *sylvestris*	r
93	*Matricaria chamomilla* L.	
94	*Medicago lupulina* L.	
95	*Medicago minima* (L.) L.	
96	*Medicago orbicularis* (L.) Bartal.	
97	*Medicago polymorpha* L.	x
98	*Medicago sativa* L.	r
99	*Mentha longifolia* (L.) Huds.	
100	*Moehringia pentandra*	r
101	*Muscari comosum* (L.) Mill.	r
102	*Oenanthe pimpinelloides* L.	
103	*Ornithogalum divergens* Boreau	
104	*Orobanche ramosa* L. subsp. *ramosa*	
105	*Ostrya carpinifolia* Scop.	
106	*Papaver rhoeas* L. subsp. *rhoeas*	r
107	*Pastinaca sativa* L. subsp. *urens* (Req. ex Godr.) Čelak.	
108	*Phleum bertolonii* DC.	
109	*Picris hieracioides* L. subsp. *hieracioides*	x
110	*Plantago lanceolata* L.	
111	*Plantago major* L. subsp. *major*	
112	*Poa bulbosa* L.	
113	*Poa pratensis* L.	
114	*Poa trivialis* L.	
115	*Polygonum aviculare* L. subsp. *aviculare*	
116	*Potentilla reptans* L.	
117	*Prunus cerasifera* Ehrh.	
118	*Pulicaria dysenterica* (L.) Bernh.	
119	*Ranunculus neapolitanus* Ten.	x
120	*Ranunculus velutinus* Ten.	
121	*Rapistrum rugosum* (L.) All.	
122	*Robinia pseudoacacia* L.	r

(continued)

Table 9.1 (continued)

n.	Species list	Cover x > 10%; r = rare
123	*Rosa canina* L. (group)	r
124	*Rubus canescens* DC. (group)	r
125	*Rumex crispus* L.	
126	*Rumex sanguineus* L.	
127	*Sanguisorba minor* Scop. subsp. *balearica* (Bourg. ex Nyman) Muñoz Garm. et C. Navarro	
128	*Schedonorus pratensis* (Huds.) P. Beauv. subsp. *apenninus* (De Not.) H. Scholz et Foggi	
129	*Sherardia arvensis* L.	
130	*Silene latifolia* Poir. subsp. *alba* (Mill.) Greuter et Burdet	
131	*Sonchus asper* (L.) Hill subsp. *asper*	
132	*Sonchus oleraceus* L.	x
133	*Stachys sylvatica* L.	
134	*Stellaria media* (L.) Vill. subsp. *media*	
135	*Sulla coronaria* (L.) Medik.	
136	*Symphytum bulbosum* K.F. Schimp.	r
137	*Tanacetum parthenium* (L.) Sch. Bip.	r
138	*Taraxacum officinale* Weber ex F.H. Wigg. (group)	
139	*Thlaspi alliaceum* L.	
140	*Thlaspi perfoliatum* L. subsp. *perfoliatum*	
141	*Thymus longicaulis* C. Presl subsp. *longicaulis*	
142	*Tilia platyphyllos* Scop. subsp. *platyphyllos*	r
143	*Tordylium apulum* L.	
144	*Torilis arvensis* (Huds.) Link subsp. *arvensis*	
145	*Tragopogon dubius* Scop.	
146	*Tragopogon porrifolius* L. subsp. *porrifolium*	
147	*Tragopogon samaritanii* Heldr. et Sartori ex Boiss.	
148	*Trifolium campestre* Schreb.	x
149	*Trifolium dubium* Sibth.	r
150	*Trifolium pratense* L. subsp. *pratense*	
151	*Trifolium repens* L. subsp. *repens*	
152	*Trifolium resupinatum* L.	x
153	*Trifolium squarrosum* L.	x
154	*Tripleurospermum inodorum* (L.) Sch. Bip.	r
155	*Urtica dioica* L. subsp. *dioica*	
156	*Verbena officinalis* L.	
157	*Veronica arvensis* L.	
158	*Veronica persica* Poir.	
159	*Vicia loiseleurii* (M. Bieb.) Litv.	
160	*Vicia sativa* L. subsp. *macrocarpa* (Moris) Arcang.	
161	*Vicia sativa* L. subsp. *nigra* (L.) Ehrh.	
162	*Vulpia ligustica* (All.) Link	

x = dominant species, r = rare species, with one or very few individuals and negligible cover

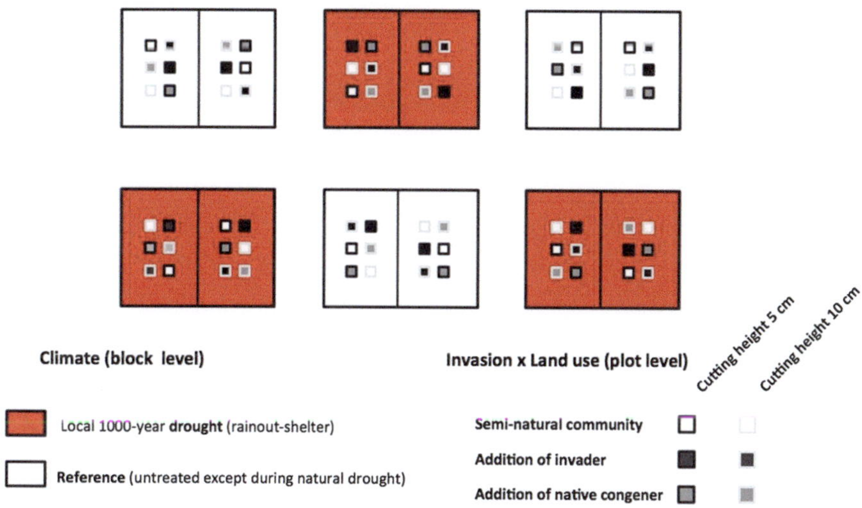

Fig. 9.2 Diagram of the experimental design, from: http://www.bayceer.uni-bayreuth.de/signal/

9.4 Main Research Questions

What is the effect of extreme weather events for grasslands in Italy and along the European gradient?

What is the role of alien species in the response of European grasslands to extreme drought events?

How does biodiversity and productivity react to drought, especially considering functional groups (i.e. legumes, graminoids, etc) and plant functional traits?

How effective are the different ways of management and mowing techniques?

9.5 The Experimental Design

The experiments of the SIGNAL research project consist of the simulation of extreme climatic events on mesic to semi-arid grasslands across ten different sites, one per nation involved (Fig. 9.2). SIGNAL focuses on the interaction of three main branches of research: biodiversity experiments, research on the impact of climate change, and the impact of invasive alien species; under different mowing regimes (Fig. 9.3).

Fig. 9.3 Covers for drought extreme events simulation; in summer

Replicating field experiments across such a varied landscape provides an excellent and unique opportunity for learning about the local degree of future grassland vulnerability in interacting global change drivers. Additionally, running the experiments under contrasting climates and with different species will allow for a previously unmatched potential for generalizations and ecological reasoning. The experiments are designed to produce regionally differentiated decision support for policy in agriculture and conservation.

In an approach to study assembly rules (Wellstein et al. 2014a, 2015), the fine-scale patterns of species combinations (rooting individuals) were sampled in 2.80 × 0.40 m blocks, subdivided into 448 micro-quadrats of 25 cm^2 (5 × 5 cm). To account for within site heterogeneity, six blocks were sampled at each site. Specific leaf area (SLA), plant height and seed mass were assessed, using literature sources and direct measurements, for all recorded vascular herbaceous species. Based on these traits, Rao's functional diversity was calculated for each micro-quadrate and compared to a null model. We used Schamp's method for randomization: keeping the abundances of species and the local species richness as in the field and assigning traits at random to each species combination. Deviation of functional diversity from random expectation was interpreted as trait divergence or trait convergence. Several occurrences of trait-based assembly rules could be detected across countries. Our site showed mostly convergence for SLA, height and seed mass. Assembly rules can be contrasting and context dependent at different grassland sites and climatic differences are often masked by local factors such as disturbance regime or soil heterogeneity.

The field experiment on the grasslands productivity also serves to assess the danger potential of two neophytes, *Lupinus polyphyllus* Lindl. and *Senecio inaequidens* DC. (Fig. 9.4). Both the alien species have been chosen because noteworthy for their weed potential in temperate meadows. They were germinated in laboratory, in Petri dishes, then planted in the experiment plots according to a common standard procedure. At the end of the experiments they will be destroyed,

Fig. 9.4 The aliens planted: a *Senecio inaequidens* sapling to the *left*; and a *Lupinus polyphyllus* seedling to the *right*

before they can disperse propagules, in order to avoid any invasions of the surrounding landscape.

The drought manipulation was conducted using differing onset dates and durations across Europe according to the local seasonality and the local 1000 year extreme drought event (see Jentsch et al. 2007 for details).

Collecting datasets from different sites may cause variability in data quality, this risk will be minimized by the use of tested protocols for the quantification of easy and clear response parameters, joint training and central laboratory analyses.

We are waiting for more insights based on further approaches studying field sampled traits (maximum height and lateral width), accounting for space occupation, and the effect of fine-scale β-diversity (at micro scale 5 × 5 cm) on resilience to drought (Wellstein et al. 2014b, c).

Acknowledgments We would like to thank all the people from the SIGNAL Consortium, for their kindness and cooperation to the elaboration and conduction of the research project; Simone Gatto, Khawla Zouglami, Francesca Netti and Mario Messini, who helped in the field works. We also thank the Camerino Montane District, owner of the area managed in agreement with University of Camerino as "Arboretum Apenninicum", and Riserva Statale Montagna di Torricchio. SIGNAL was funded by the ERA-Net BiodivERsA, with the national funders Belgian Science Policy Office (belspo), German Federal Ministry of Education and Research (BMBF), Bulgarian Science Found and Ministère de l'Écologie, du Développement durable et de l'Énergie (France) as parties of the 2011–2012 call for research proposals.

References

Campetella G, Cianfaglione K, Wellstein C, Chelli S, Cervellini M, Canullo R, Consorzio SIGNAL (2013) Progetto signal: gradienti europei di resilienza rispetto ad eventi climatici estremi. XXIII Congresso della Società Italiana di Ecologia (SItE). 16–18/9/2013, Ancona, Italy. Poster

Cianfaglione K (2014) On the potential of Quercus pubescens Willd. and other species of Quercus in the Camerino syncline (Central Italy). In: Box EO, Kazue F (eds) Warm-temperate deciduous forests around the Northern Hemisphere. Springer, Switzerland, pp 165–174

Cianfaglione K, Wellstein C, Chelli S, Campetella G, Cervellini M, Canullo R (2013) Progetto SIGNAL: effetti di eventi climatici estremi sulle praterie europee. 108° Congresso Società Botanica Italiana Onlus (SBI). 18–20/9/2013, Baselga di Piné (TN), Italy. Poster

De Santo RS (1978) Concepts of applied ecology, Heidelberg science library. Springer, New York

Díaz S, Cabido M (2001) Vive la différence: plant functional diversity matters to ecosystem processes. Trends Ecol Evol 16(11):646–655

Gorfer A (1982) I segni della storia. Saturnia ed. Trento, Italy.

Jentsch A, Kreyling J, Beierkuhnlein C (2007) A new generation of climate change experiments: events, not trends. Front Ecol Environ 5:365–374

Jentsch A, Kreyling J, Apostolova I, Bahn M, Bartha S, Beierkuhnlein C, Bloor J, de Boeck H, Dengler J, Picon-Cochard C, Campetella G, Canullo R, Nijs I, Stampfli A, Sternberg M, Uğurlu E, Walter J, Wellstein C, Zeitler M, SIGNAL PhD students (2014) Joining biodiversity experiments, climate change research and invasion biology to assess European gradients of grassland resilience in the face of climate extremes. In: Mucina L, Price JN, Kalwij JM (eds) Biodiversity and vegetation: patterns, processes, conservation. 57th IAVS Symposium. Kwongan Foundation, Perth, p 114

Leadley PW, Pereira HM, Alkemade R, Fernandez-Manjarrés J, Proença V, Scharlemann JPW, Walpole M (eds) (2010) Biodiversity scenarios: projections of 21st century change in biodiversity and associated ecosystem services. Secretariat of the Convention on Biological Diversity, Montreal. CBD Technical Series No. 50

Maynard-Smith J (1974) Models in ecology. University press, Cambridge

Pedrotti F (1992) Il progetto per l'istituzione dell'arboreto dell'Università di Camerino "Arboretum Apenninicum". L'Uomo e l'Ambiente (Camerino) 15:7–16

Spehn EM, Scherer-Lorenzen M, Schmid B, Hector A, Caldeira MC, Dimitrakopoulos PG, Finn JA, Jumpponen A, O'Donnovan G, Pereira JS, Schulze E-D, Troumbis AY, Körner C (2002) The role of legumes as a component of biodiversity in a cross-European study of grassland biomass nitrogen. Oikos 98:205–218

Wellstein C, Campetella G, Spada F, Chelli S, Mucina L, Canullo R, Bartha S (2014a) Context-dependent assembly rules and the role of dominating grasses in semi-natural abandoned sub-Mediterranean grasslands. Agric Ecosyst Environ 182:113–122

Wellstein C, Jentsch A, Chelli S, Campetella G, Canullo R, Apostolova I, Bloor J, Cianfaglione K, Dengler J, Von Gillhaußen P, Güler B, Házi J, Komoly C, Kreyling J, Pottier J, Szabó G, Terziiska T, Uğurlu E, Zimmermann Z, Bartha S (2014b) Finescale functional organization in grasslands: a microquadrat-based approach to assembly rules (Gyepek funkcionális szerveződése mikrocönológiai megközelítésben). International conference "II. Sustainable development in the Carpathian Basin", University of Debrecen, Hungarian Biological Society, Budapest Zoo and Botanical Garden. 11–12/12/2014, Budapest, Hungary. Poster

Wellstein C, Jentsch A, Chelli S, Campetella G, Canullo R, Apostolova I, Bloor J, Cianfaglione K, Dengler J, von Gillhaußen P, Guler B, Hazi J, Komoly C, Kreyling J, Pottier J, Szabo G, Terziiska T, Uğurlu E, Zimmermann Z, Bartha S (2014c) Trait-based assembly rules across climatic gradients of European grasslands. In: Mucina L, Price JN, Kalwij JM (eds) Biodiversity and vegetation: patterns, processes, conservation. 57th IAVS Symposium. Kwongan Foundation, Perth, p 266

Wellstein C, Jentsch A, Chelli S, Campetella G, Canullo R, Apostolova I, Bloor J, Cianfaglione K, Dengler J, von Gillhaußen P, Güler B, Házi J, Komoly C, Kreyling J, Pottier J, Sopotlieva D, Szabó G, Terziiska T, Uğurlu E, Zimmermann Z, Bartha S (2015) Trait-based assembly rules across climatic gradients of European grasslands are affected little by extreme drought. In: Chytrý M, Zelený D, Hettenbergerová E (eds), Understanding broad-scale vegetation patterns. 58th Annual Symposium of the International Association for Vegetation Science, 19–24 July 2015, Masaryk University, Brno, Czech Republic. Abstracts: 406

Yachi S, Loreau M (1999) Biodiversity and ecosystem productivity in a fluctuating environment: the insurance hypothesis. Proc Natl Acad Sci 96(February):1463–1468

An Altitudinal Gradient for Breeding Birds in Marche Region (Central Italy) 10

Paolo Forconi

Abstract

Species distribution is determined by several factors. As of 2013, the study area in central Italy, hosts 145 breeding birds. To explore the relationship between species richness and elevation, the study area was divided into 4 altitudinal zones, 600 m wide and 4 belts were indicated for the potential vegetation gradient, as these belts provide a better biological sense as natural marker of climatic and altitudinal transitions. Plotting the number of species by altitudinal belts revealed that species diversity declines in correspondence with increasing elevation. Thus, the lowest altitudinal belt, hilly, has the highest richness in species, (120 sp.), followed by low mountain (110), mountain (80) and high mountain (14). Instead, analysis according to potential vegetation belts indicates that the highest species richness is found in the second lowest belt: coastal (95 sp.), hilly (128), mountain (93) and high mountain (14). The same distribution is obtained for the exclusive species in each vegetation belt (5, 12, 5, 2 respectively) and for birds in Annex I of Birds Directive (2009/147/CE).

From the point of view of conservation, we have to consider not only breeding birds, but also wintering and migrating birds. In fact the distribution map of birds of european conservation concern (breeding, wintering and migrating) emphasizes the highly important role of the coastal area followed by mountain and main rivers connecting them. It is possible to obtain different outcomes as we apply different methods. Potential vegetation belts seem to describe species distribution better than do altitudinal belts. Moreover, potential vegetation belts have more sense than arbitrary altitudinal belts. From the point of view of conservation, considering only breeding birds may provide misleading data. In a planning process, all bird species should be considered: breeding, migrating and wintering.

P. Forconi (✉)
Studio Faunistico Chiros, Macerata, Italy
e-mail: paolo.forconi@libero.it

F. Pedrotti (ed.), *Climate Gradients and Biodiversity in Mountains of Italy*, Geobotany Studies, https://doi.org/10.1007/978-3-319-67967-9_10

Keywords
Bird gradient • Altitudinal belts • Potential vegetation belts • Breeding birds

10.1 Introduction

Species distribution is determined by several factors: climate (temperature, humidity, solar radiation), productivity, ecotones, competition, biogeography, area extension, isolation, anthropogenic alteration, habitat structure and heterogeneity (Terborgh 1977, 1985; Ricklefs and Schluter 1993; Huston 1999; Brown 2001; Lomolino 2001; Kattan and Franco 2004; Joshi et al. 2012).

Mountain ecosystems are usually recognized as biodiversity hotspots, often with high number of endemic species. In mountains the decrease in species richness with increasing elevation is a widely recognized pattern (Begon et al. 1990; Lomolino 2001). Others suggest a unimodal relationship with a peak in species richness at low to mid elevations (Rahbek 1995; Stotz et al. 1996; Stotz 1998; Brown 2001; Lomolino 2001; Kattan and Franco 2004; Gaston and Spicer 2004).

This study aims to evaluate the distribution of the diversity of breeding birds in the Marche region (central Italy) in relation to altitudinal and vegetational gradients.

10.2 Study Area

Marche region (43°N, 13°E) is extended for 9382.82 km^2, arising from costline of Adriatic Sea to hills and mountains up to 2476 m asl. The potential vegetation can be divided into the following four categories (Pedrotti et al. 1981 with updates; Blasi 2010):

1. The coastal strip, represented by beaches or rocky shores (Fig. 10.1) with dunes and maquis. This belt is very small and it is considerably transformed because of the presence of urban, industrial, agricultural areas and intense tourism exploitation. Almost nothing remains of the original dunes and wetlands. Some artificial lakes were created as a result of quarrying or for hunting purposes.

   Main Orders are: *Ammophiletalia* (vegetation of the sand dunes) and *Quercetalia ilicis* (maquis).
2. The hilly area was also significantly transformed by man into cultivated fields. The wooded areas are very small and isolated as well as the riparian vegetation along rivers is limited to a few meters from the shore (Fig. 10.2), except in the steeper areas (Fig. 10.3).

   Main Orders are: *Quercetalia-pubescenti petraeae* (thermophilic deciduous forests), *Salicetalia albae* (willow of water courses) and *Populetalia albae* (poplar groves).

Fig. 10.1 The coastal belt in Marche region (Conero Regional Park)

Fig. 10.2 The hilly belt in foreground and the mountain and subalpine-alpine belts in background

Fig. 10.3 The connection between the hilly and the mountain belts

Fig. 10.4 The mountain and the subalpine-alpine belts (Sibillini National Park)

3. The mountain belt (Fig. 10.4) was converted mainly with the creation of secondary grasslands, but there are still many wooded areas although of poor quality due to frequent cuts that do not allow the development of large trees. The main Order is *Fagetalia sylvaticae* (deciduous forests of beech).
4. The subalpine and alpine belt (Fig. 10.4) is the one that has the highest level of wilderness, with the presence of primary grasslands. However, the belt of twisted shrubs (1800–2000 m asl) was also eliminated in the past with the creation of pastures for sheep.

Main Orders are: *Vaccinio-Piceetalia* (subalpine heaths), *Prunetalia spinosae* (subalpine shrubs), *Seslerietalia apenninae* (primary grasslands).

Anthropic transformation have led to a general impoverishment and extinction of some specialised birds, but also an expansion of other species linked to anthropic and open environments.

10.3 Methods

Bird species list was compiled with elevation and vegetation ranges. Data were obtained from Giacchini (2003), Giacchini et al. (2013) and personal database.

All land and freshwater species were considered, for 145 breeding species included in the analysis, updated to 2013.

To explore the relationship between species richness and elevation, because belts should have the same amplitude to be comparable, the study area was divided in four altitudinal zones, 600 m wide between 0 and 2400 m: hilly (0–600 m), low mountain (600–1200 m), mountain (1200–1800 m) and high mountain (1800–2400 m) (Fig. 10.5).

Nevertheless these belts are arbitrary while the potential vegetation (described in the Study area) has a better biological sense as natural marker of climatic and altitudinal transitions.

For the potential vegetation gradient, four belts have been identified: coastal (mediterranean maquis), hilly (oak forests—up to about 1000 m above sea level), mountain (beech forest—from 1000 to 1800 m) and subalpine-alpine (altitude primary grasslands >1800 m) (Fig. 10.6).

The area of each belt was calculated using QGis.

The Species Density Index indicate the number of species every 100 kmq, and was calculated putting in relation the species in every belts with the corresponding surface.

To relate species distribution with conservation, the species of European Conservation Concern (Directive 2009/147/CE) as breeding, wintering and migrating were mapped on a grid of 10 × 10 km.

A value was attributed to every species corresponding to:

- 0.5 for migration
- 1 for resting during migration or presence
- 2 for breeding.

Double values were attributed to prioritary species for Europe.

Applying kernel density with Spatial Analyst, the distribution of high density areas was obtained.

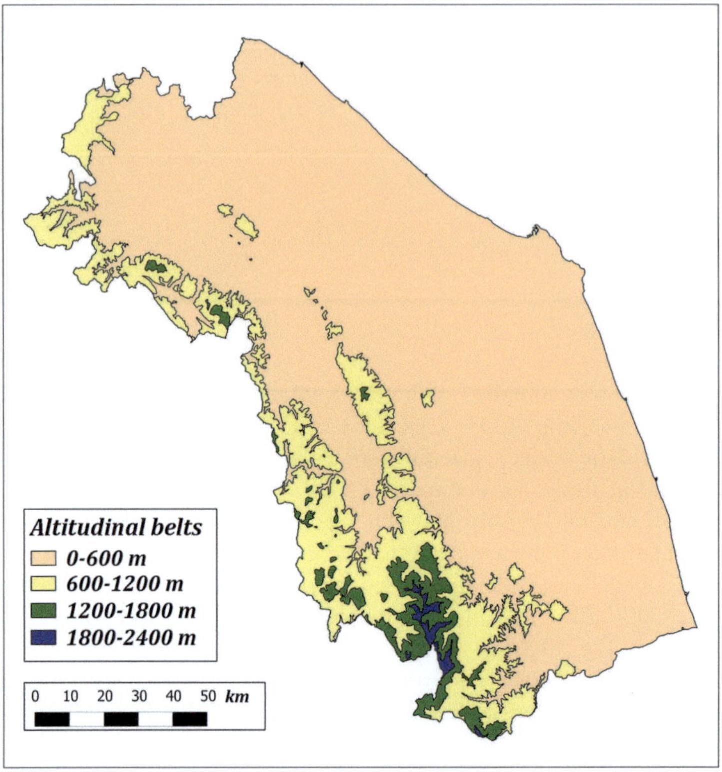

Fig. 10.5 The study area with altitudinal belts

10.4 Results

Plotting the number of species by altitudinal belts revealed a decline in species diversity with increasing elevation (Fig. 10.7) with lowest belt reaching the highest richness in species: hilly (120 sp.), low mountain (110), mountain (80) and high mountain (14).

Instead, by potential vegetation belts there is an highest species richness in the second lowest belt (Fig. 10.8): coastal (95 sp.), hilly (128), mountain (93) and subalpine-alpine (14).

The same distribution is obtained for the exclusive species in each vegetation belts.

The coastal belt has five exclusive species: Red-crested Pochard (*Netta Rufina*), Kentish Plover (*Charadrius alexandrinus*), Herring Gull (*Larus michahellis*), Pallid Swift (*Apus pallidus*) and Short-toed Lark (*Calandrella brachydactyla*).

The exclusive species of the hilly belt are 12 including: Pygmy Cormorant (*Phalacrocorax pygmeus*), Red Kite (*Milvus milvus*), Short-toed Eagle (*Circaetus

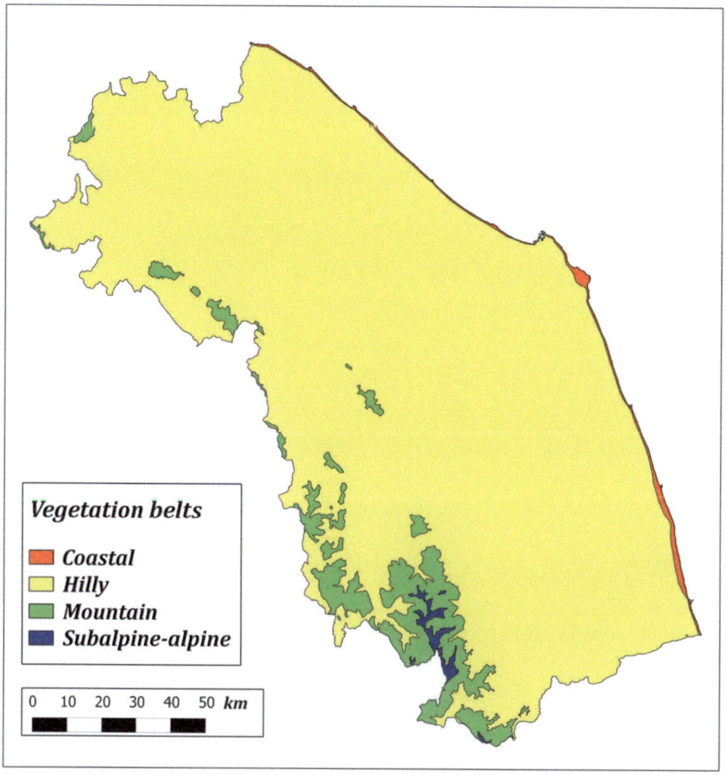

Fig. 10.6 The study area with potential vegetation belts

gallicus), Montagu's Harrier (*Circus pygargus*), Lanner (*Falco biarmicus*) and Raven (*Corvus corax*).

The exclusive species of the mountain belt are five including: Wood Warbler (*Phylloscopus sibilatrix*), Collared Flycatcher (*Ficedula albicollis*) and Eurasian Treecreeper (*Certhia familiaris*).

The exclusive species of the high mountain are two: Alpine Accentor (*Prunella collaris*) and White-winged Snowfinch (*Montifringilla nivalis*).

Nevertheless, surface area is very different among belts (Tables 10.1 and 10.2) and there is no correlation between area and number of species. Considering surface area in each altitudinal belt, the Species Density Index revealed higher values for higher altitudes, while in potential vegetation belts the highest value is for coastal belt (Tables 10.1 and 10.2) (Fig. 10.9). This latter is masked in altitudinal belts by contiguous hilly belt.

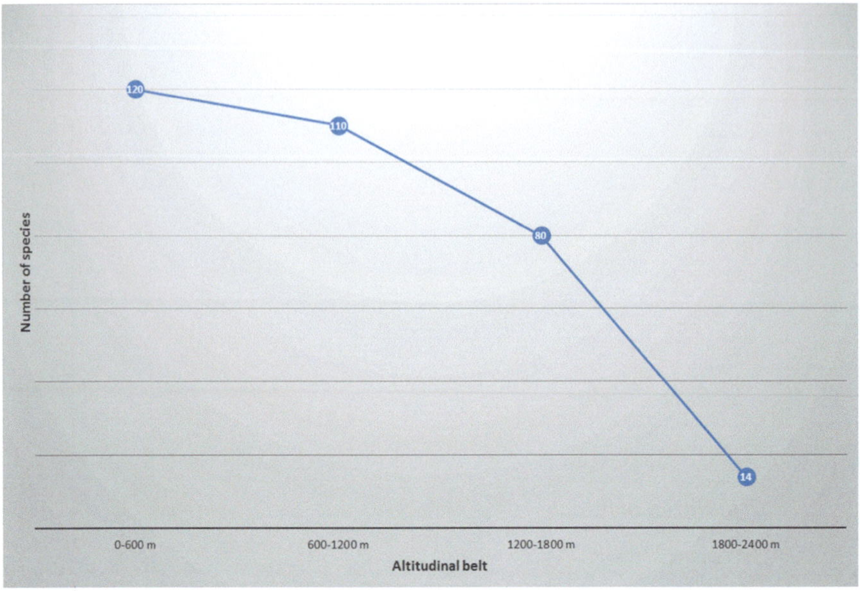

Fig. 10.7 Relationship between breeding species richness and altitudinal belts in Marche region

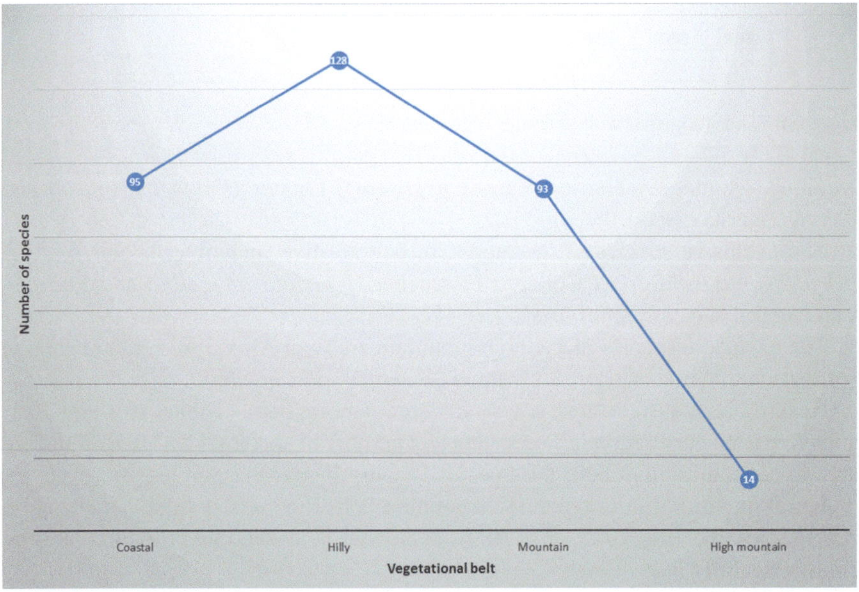

Fig. 10.8 Relationship between breeding species richness and potential vegetation belts in Marche region

Table 10.1 Breeding species richness and altitudinal belts in Marche region

Altitudinal belts	Area (km^2)	% area	N. species	% species	Species/100 km^2
0–600	7431.07	79.20	120	82.76	1.61
600–1200	1605.39	17.11	110	75.86	6.85
1200–1800	303.66	3.24	80	55.17	26.35
1800–2400	42.70	0.46	14	9.66	32.79
Area Marche region	9382.82	100			

Table 10.2 Breeding species richness and potential vegetation belts in Marche region

Vegetational belts	Area (km^2)	% Area	N. species	% species	Species/100 km^2
Coastal	79.44	0.85	95	65.52	119.59
Hilly	8676.03	92.47	128	88.28	1.48
Mountain	584.65	6.23	93	64.14	15.91
Subalpine-Alpine	42.70	0.46	14	9.66	32.79
Area Marche region	9382.82	100			

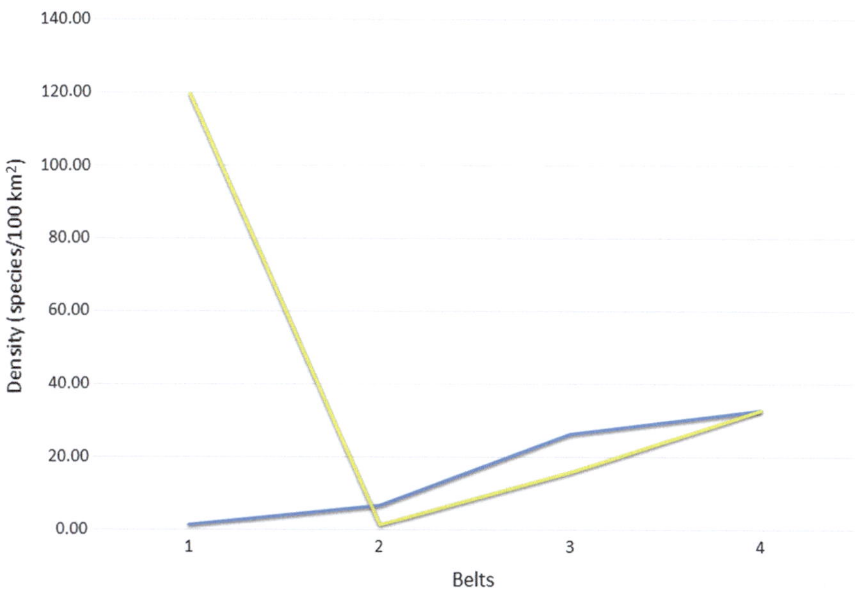

Fig. 10.9 Breeding species density index/100 km^2 for altitudinal belts (*blue*) and potential vegetation belts (*yellow*)

10.4.1 Species of European Conservation Concern

Considering only birds in Annex I of Birds Directive (2009/147/CE), they have the same distribution of the overall breeding birds (Fig. 10.10).

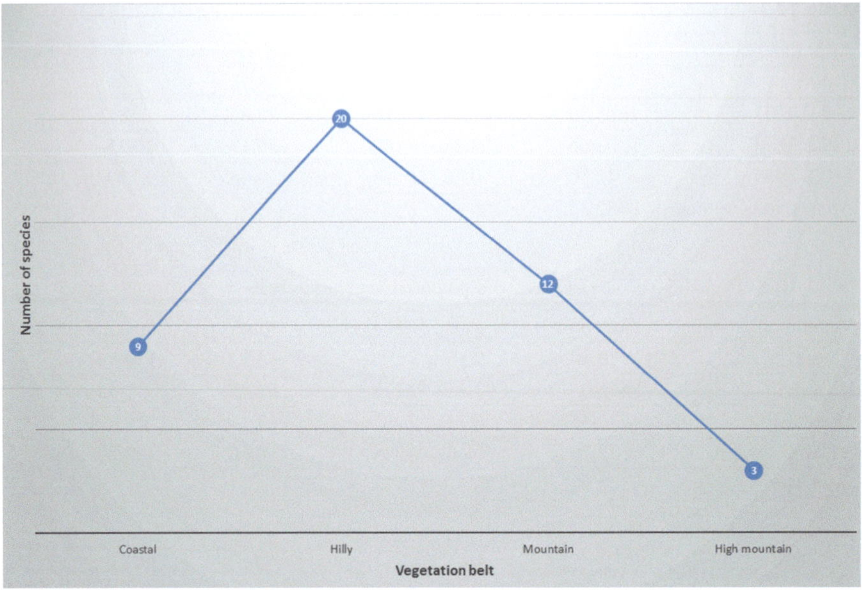

Fig. 10.10 Breeding species of European conservation concern (Annex I in Birds Directive 2009/147/CEE)

From the point of view of conservation, we have to consider not only breeding birds, but also wintering and migrating birds.

The map in Fig. 10.11 shows the distribution of birds of european concern (breeding, wintering and migrating) and emphasize the high important role of the coastal area followed by mountain and main rivers connecting them. Infact wintering and migrating birds mainly use coastal areas.

10.5 Discussion

The results show that it is possible to obtain different outcomes as we apply different methods.

Potential vegetation belts seem to describe better species distribution than altitudinal belts. Moreover, potential vegetation belts have more sense than arbitrary altitudinal belts.

From the point of view of conservation, considering only breeding birds may provide misleading data. In a planning process, all birds species should be considered, breeding, migrating and wintering.

Marche region presents some weaknesses in the protection of coastal and hilly areas, while mountain areas are well protected by Natura 2000 sites (Forconi et al. 2005).

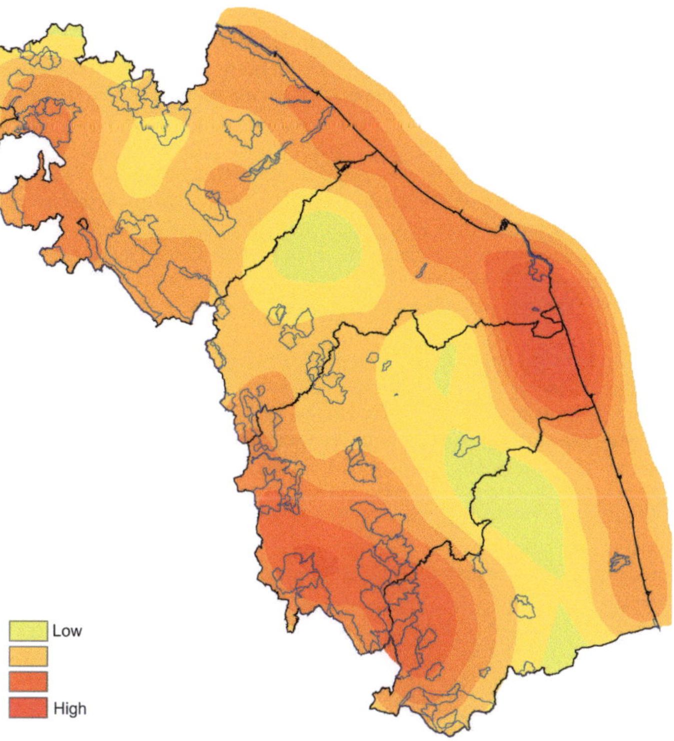

Fig. 10.11 Distribution of breeding, wintering and migrating birds of European conservation concern in Marche region. The *blue areas* are SAC and SPA as indicated by Birds and Habitat Directive

In addition, exploitation by man, and the resulting anthropogenic environmental changes, has greatly reduced species richness in Marche region, particularly in the middle and lowest belts, where the human presence is greater.

References

Begon M, Harper JL, Townsend CR (1990) Ecology: individuals, populations, communities. Blackwell Scientific Publications, London, pp 203–325

Blasi C (a cura di) (2010) La vegetazione d'Italia. Palumbi ed, Roma

Brown JH (2001) Mammals on mountainsides: elevational patterns of diversity. Glob Ecol Biogeogr 10:101–109

Forconi P, Polini N, Fusari M, Marini G, Di Martino V (2005) I siti Natura 2000 nella regione Marche: situazione attuale e necessità di aggiornamento. Avocetta 29:111

Gaston KJ, Spicer JI (2004) Biodiversity: an introduction, 2nd edn. Blackwell, Oxford

Giacchini P (2003) Check-list degli uccelli delle Marche. Riv Ital Orn 73(1):25–45

Giacchini P, Forconi P, Fusari M, Gambelli P, Marini G, Mencarelli M, Morganti N, Morici F, Pascucci M, Polini N, Pruscini F (2013) Aggiornamento dell'avifauna nidificante nelle Marche. Atti XVII Convegno Italiano di Ornitologia, Trieste 11–15 settembre 2013

Huston MA (1999) Local processes and regional patterns: appropriate scales for understanding variation in the diversity of plants and animals. Oikos 86:393–401

Joshi KK, Bhatt D, Thapliyal A (2012) Avian diversity and its association with vegetation structure in different elevational zones of Nainital district (Western Himalayan) of Uttarakhand. Int J Biodivers Conserv 4(11):364–376

Kattan GH, Franco P (2004) Bird diversity along elevation gradients in the Andes of Colombia. J Biogeogr 13:451–458

Lomolino MV (2001) Elevation gradients of species density historical and prospective views. Glob Ecol Biogeogr 10:3–13

Pedrotti F, Ballelli S, Biondi E, Cortini PC, Francalancia C, Orsomando E (1981) Carta del paesaggio vegetale delle Marche. Università di Camerino, Regione Marche

Rahbek C (1995) The elevational gradient of species richness: a uniform pattern? Ecography 18:200–205

Ricklefs RE, Schluter D (1993) Species diversity: regional and historical influences. In: Ricklefs RE, Schluter D (eds) Species diversity in ecological communities: historical and geographical perspective. University of Chicago, Cambridge, pp 350–363

Stotz DF (1998) Endemism and species turnover with elevation in montane avifauna in neotropics: implications for conservation. In: Mace GM, Balmford A, Ginsberg JR (eds) Conservation in a changing world. Cambridge University, New York, pp 161–180

Stotz DF, Fitzpatrick JW, Parker TA, Moskovits DK (1996) Neotropical birds: ecology and conservation. Chicago University, Chicago

Terborgh J (1977) Bird species diversity on an Andean elevational gradient. Ecology 58:1007–1019

Terborgh J (1985) The role of ecotones in the distribution of Andean birds. Ecology 66:1237–1246

MIX
Papier aus verantwortungsvollen Quellen
Paper from responsible sources
FSC® C105338

If you have any concerns about our products,
you can contact us on
ProductSafety@springernature.com

In case Publisher is established outside the EU,
the EU authorized representative is:
Springer Nature Customer Service Center GmbH
Europaplatz 3, 69115 Heidelberg, Germany

Printed by Libri Plureos GmbH
in Hamburg, Germany